AF526352

Parallel and Distributed Discrete Event Simulation

PARALLEL AND DISTRIBUTED DISCRETE EVENT SIMULATION

CARL TROPPER
EDITOR

Nova Science Publishers, Inc.
New York

Senior Editors: Susan Boriotti and Donna Dennis
Coordinating Editor: Tatiana Shohov
Office Manager: Annette Hellinger
Graphics: Wanda Serrano
Editorial Production: Jennifer Vogt, Matthew Kozlowski, Jonathan Rose, Ron Doda and Maya Columbus
Circulation: Ave Maria Gonzalez, Indah Becker and Vladimir Klestov
Communications and Acquisitions: Serge P. Shohov
Marketing: Cathy DeGregory

Library of Congress Cataloging-in-Publication Data
Available upon request.
ISBN 1-59033-377-2

400 Oser Ave, Suite 1600
Hauppauge, New York 11788-3619
Tele. 631-231-7269 Fax 631-231-8175
e-mail: Novascience@earthlink.net
Web Site: http://www.novapubishers.com

Printed in the United States of America

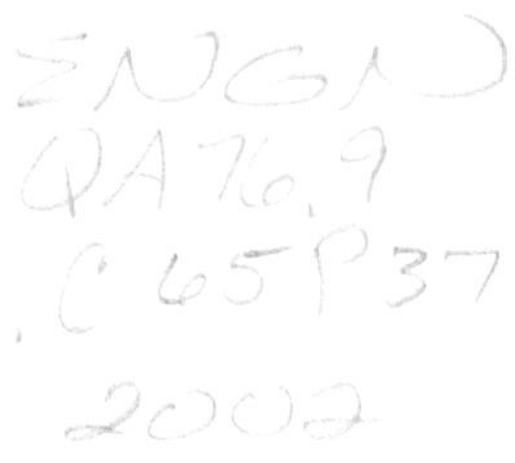

CONTENTS

APPLICATIONS OF PARALLEL AND DISTRIBUTED DISCRETE EVENT SIMULATION

CARL TROPPER

Discrete-event simulation has long been an integral part of the design process of complex engineering systems and the modelling of natural phenomena. Many of the systems which we seek to understand or control can be modelled as digital systems. In a digital model, we view the system at discrete instants of time, in effect taking snapshots of the system at these intstants. For example, in a computer network simulation an event can be the sending of a message from one node to another node while in a VLSI logic simulation, the arrival of a signal at a gate may be viewed as an event.

Each event in a discrete-event simulation has a timestamp associated with it. When an event is processed, it is possible that new events are generated as a consequence of this processing. These new events have larger timestamps, obtained by adding a simulation time advance to the timestamp of the event which it had prior to processing. The events in the simulation are stored in a heap and are processed in order of the lowest timestamp first.

Digital systems such as computer systems are naturally susceptible to this approach. However, a variety of other systems may also be modelled this way. These inclued transportation systems such as air-traffic control systems, epidemological models such as the spreading of a virus, and military war-gaming models.

As the systems and phenomena we want to model increase in size and complexity, the memory demands of these simulations increase and it becomes increasingly difficult to obtain acceptible execution times. The circuits which we want to simulate now contain hundreds of millions of gates. A detailed simulation of the Internet would make inordinate demands on a workstation. In order to accomodate the grwoing need for the simulation of larger models, it became necessary to make use of distributed and parallel computer systems to execute the simulations. In the early 90's parallel machines were made use of while more recently the use of clusters of workstations (Beowulf, Myrinet) are utilized as they represent a more cost-effective approach then parallel machines. In addition, shared memory multiprocessors are now much more cost-effective.

Like any other distributed program, a distributed simulation is composed of processes which communicate with one another. The communication may occur via shared memory or via message passing. The processes in a distributed simulation are each intended to simulate a portion of the system being modelled and are referred to as Logical Processes (LPs). An LP creates events, sends events to other LPs and receives events from other LPs in the course of a simulation. Associated with each LP are input queues used to store messages from other LPs.

The advance of time in a distributed simulation poses an intriguing problem because of its inherant lack of global memory. The events of a distributed simulation are spread among the processors and consequently time is advanced in each process independantly of the other processes. This is accomplished via the notion of Local Simulation Time (LST) which is maintained by each LP. When an event is processed at an LP, the LST takes on the value of its timestamp. The LP processes events from its input queues by selecting the event bearing the smallest timestamp from all of its queues.

The treatment of time in a distributed system is of fundamental importance to the understanding and building of distributed systems. [16] contains a discussion of the nature of time in a distributed system for the interested reader. It was necessary to develop synchyronization strategies for the distributed and parallel programs which execute distributed simulations. As mentioned before, in a uniprocessor events are stored in a heap and simulated in the order of the smallest timestamp first. Since the events of a distributed simulation are spread among the processors it is possible to execute events out of their correct order. This happens if an event with a smaller timestamp then an event which has already been processed arrives at a processor from another processor. In a military simulation, it matters in which order the events "aim the gun" and "fire the gun" are executed. The central problem of Distributed simulation is the development of synchronization algorithms which are capable of maintaining causality and which do so with minimal overhead. The interested reader should consult the proceedings of the IEEE Workshop on Parallel and Distributed Simulation (PADS) and a recent book which describes the field [10].

Two primary approaches to synchronization algorithms have been developed, the *conservative* and *optimistic* classes of algorithms. A conservative algorithm is characterized by its blocking behavior. In a conservative simulation, if one of the input queues at an LP is empty the LP blocks awaiting a message from another LP. This behavior exacts a price, however, in the form of increased execution time and the possible formation of deadlocks. A deadlock forms if a group of LPs is arranged in the form of a cycle such that each of the LPs is awaiting a message from its predecessor in the cycle. More generally, a deadlock occurs if the LPs are arranged in the form of a knot. Hence it becomes necessary to either find means to avoid deadlocks or to detect and break them. There are a plethora of algorithms for each approach. We describe several algorithms.

In the null message approach [9], each time an LP sends a message to a neighboring LP, it also sends a message (the null message) to its other neighbors containing the timestamp of the message, thereby causing the neighbors to advance their LST's. As a consequence, all of the LPs are aware of the earliest simulation time that a message can arrive at an empty input queue. The LP can use this information to avoid blocking by comparing the smallest timestamp in each of its input queues to this value. If the smallest timestamp in all of the input queues is samller then this value then it can process the associated event. Otherwise it must block. The drawback of this approach is clearly the large number of messages which must be sent; as a consequence much work has been done to avoid the sending of a large number of null messages. [13] and [14] are efforts in this direction. Another form of deadlock avoidance is a time window approach in which tiem windows are successively generated with the property that the events contained in these windows are safe to process [1].

The deadlock detection and breaking approach relies upon algorithms to detect deadlocks. For example, in [19], a knot detection algorithm is used to detect a deadlock, and the event bearing the minimal timestamp in the knot is detected as well. This event is safe to execute.

The other major class of synchroniztion algorithm is known as optimistic. Time Warp [4] is the prime example in this category. In optimistic algorithms LPs maintain one input queue and all of the events which arrive from other LPs are stored in the queue. The events are processed without any concern for the arrival of events with smaller timestamps. If such a "straggler" eventarrives at an LP, the LP restores its state just prior to the arrival of the straggler and continues with the simulation from that point, a process known as rolling back. The LP must maintain checkpoints of previous states in order to roll back. In addition, it is necessary to to cancel messages which were produced subsequent to the straggler as they may well be incorrect. In order to do this, each LP maintains an output queue in which it stores copies of messages which it has already sent. Upon the arrival of a straggler, the LP sends these copies to the same LPs which received the original messages. If the message and its copy meet in an input queue, the two messages cancel (anihilate) one another. For this reason, the copy is known as an anti-message. If the anti-message arrives after the original message has been processed, the destination LP rolls back to the time of the anti-message and sends out its own anti-messages. Clearly, the memory demands imposed by storing copies of LP's states and maintaining anti-messages in the output queue are onerous. Hence techniques to reduce the amoutn of memory used by Time Warp are fundamental to its success. One important technique involve the computation of the smallest simulation time to which any LP in the simulation may roll back, known as the Global Virtual Time (GVT). If we define the LVT as the timestamp of the last message processed at an LP, then the GVT is the minimum of (1) the LVT values of all of the LPs in the simulation and (2) the minimum timestamp of all of the events which have been sent but which have not been processed at a given point in real time. Since no LP can roll back prior to the GVT, all of the memory allocated prior tothe GVT may be released. Many algorithms for computing the GVT have been developed [7]. Another technique to reduce the amount of memory employed is to periodically checkpoint the states of an LP, instead of after every event.

In a shared memory multiprocessor the use of direct cancellation [12] is used to control the use of memory. Here pointers are used to link the descendants of an event to the event, thereby eliminating the need for anti-messages. Recently, a direct cancellation technique for distributed memroy environments has been developed [17]. Memory management techniques are also used to reclaim space from LPs so that a stalled simulation may be allowed to continue [5, 6, 12].

In recent years the emphasis in the field has changed from developing efficient synchronization techniques to the application of these techniques to real world problems. In the area of computer networks, an on-going effort is simulation of the Internet [8]. The intention of the project is to provide a realistic test-bed for the development of new Internet protocols and to locate performance problems. A detailed distributed simulation of the Internet provides an environment in which experimental conditions can be controlled and in which it is possible to replicate experiments. To date, it has been necessary to experiment on the actual network itself.

The simulation of VLSI circuitry provides another fruitful area for the application of distributed simulation. A simulation environment for VHDL circuitry is described in [15]. A similar project for Verilog is underway.

Military simulations are benefitting from advances in distributed simulation. An important application is the uniting of existing simulations of different aspects of combat together, e.g. tank warfare, close air-support and infantry warfare. The inclusion of live exercises into this environment is also being pursued. The field of distributed interactive simulation (DIS) is devoted to these advances. A number of conferences are devoted to this area, among them the IEEE Real Time and Distributed Interactive simulation conference.

Yet another area is the simulation of large manufacturing environments, such as the production of VLSI wafers.

The chapters in this book are representative of the advances in these fields. In "The Development of Conservative Superstep Protocols for Shared Memory Systems", Gan et al provides strong evidence for the utility of (conservative) distributed simulation of a wafer fabrication process. The performance of several different synchronous conservative protocols are compared on a number of models based on data sets supplied by Sematech. Sematech is a consortium of semiconductor manufacturing companies that does research for its members in semiconductor manufacturing. In the "Implementation of a Virtual Time Synchronizer for Distributed Databases on a Cluster of Workstations", Boukerche et al study the performance of a distributed database system implemented on a network of workstations and synchronized by an optimistic protocol. In "Applying Multilevel Paritioning to Parallel Logic Simulation", Subramaninan et al describe a partitioning algorithm for logic simulation and examine its performance making use of the optimistic simulation kernel which lies at the heart of the VHDL simulator referred to above. In "Self-Organizing Criticality in Optimistic Simulation of Correlated Systems". Overeinder et al examine the relationship between the rollback behavior in Time Warp and hte physical complexity of Ising Spin systems. Ising Spin models are used to simulate the magnetization of ferro-metals. Finally, Moradi et al study the DOD's High Level Architecture (HLA) in the context of a simulation of an air-traffic control system.

REFERENCES

[1] TURNER, S. AND XU, M., *Performance Evaluation of the Bounded Time Warp Algorithm*, Proceedings of the SCS Multi-conference on PADS, volume 22, pages 117-126, 1992.

[2] DAS, S. AND FUJIMOTO, R., *An Adaptive Memory Management Protocol for Time Warp Parallel Simulation*, Proceedings of ACM SIGMETRICS Conference on Measure-

ment and Modeling of Computer Systems, volume 22, pages 201-210, 1994.

[3] R. FUJIMOTO, *Time Warp on a Shared Memory Multiprocessor*, Transactions of the Society for Computer Simulation, Vol. 6, No. 3, pp. 211-239, July 1989.

[4] D.A. JEFFERSON, *Virtual Time*, ACM Transactions on Programming Languages and systems, Vol. 7, No. 3, pp. 404-425, July 1985.

[5] D.A. JEFFERSON, *Virtual Time II: The Cancelback Protocol for Storage Management in Time Warp*, Proc. 9th Annual ACM Symposium on Principles of Distributed Computing, pp 75-90, ACM, 1990

[6] Y-B LIN, E. LAZOWSKA, *Reducing the State Saving Overhead for Time Warp Parallel Simulation*, T-R 90-02-03, Dept. of Computer Science and Engineering, Univ. Washington, Seattle, WA, 1990

[7] MATTERN, F., *Efficient Algorithms for Distributed Snapshots and Global Virtual Time Approximation*, Journal of Parallel and Distributed Computing, 18: 423-434, 1993.

[8] D. NICOL, J. CROWE, A. OGIELSKI, *Modelling of the Global Internet*, Computer Science and Engineering, vol1, no1, Jan-Feb 1999, pp. 42-50.

[9] K.CHANDY, J.MISRA, *Distributed Simulation:A Case Study in the Design and Verification of Distriuted Programs*, IEEE Trans. Software Eng S-5, Sept. 1979, pp. 440-452

[10] R. FUJIMOTO, *Parallel and Distributed Simulation Systems*, Wiley Interscience, 2000

[11] H. AVRIL, C. TROPPER, *Clustered Time Warp and Logic Simulation*, Proceedings of the 9th Workshop on Parallel and Distributed Simulation,1995, pp112-119.

[12] S. DAS, R. FUJIMOTO,K. PANESAR, D. ALLISON, M. HYBINETTE *GTW: A Time Warp System for Shared Memory Multiprocessors* Proceedings of the 1994 Winter Simulation Conference, 1994

[13] J. MISRA, *Distributed discrete event simulation*, ACM Computing Survey 18,1 ,March 1986, pp 39-65

[14] W. CAI, E. LETERTRE, S.J. TURNER,*Dag Consistent Parallel Simulation: a Predictable and Robust Conservative Algorithm*, Proc. 11th Workshop on Parallel and Distributed Simulation (PADS97), pp 178-181, Lockenhaus, Austria, June 1997

[15] P. WILSEY,ET AL, *Analysis and Simulation of Mixed Technology VLSI Systems*, Special Issue of Journal of Parallel and Distributed Computing, to appear, April 2002

[16] L. LAMPORT, *Time, Clocks, and the ordering of events in a distributed system*, Communications of the ACM, vol21(7), pp. 558-565.

[17] J-L. ZHAO, C. TROPPER, *The Dependence List in Time Warp*, Proc. Workshop on Parallel and Distributed Simulation (PADS01), to appear, Los Angeles, California, May 2001

[18] D. E. MARTIN, R. RADHAKRISHNAN, D. M. RAO, M. CHETLUR, K. SUBRAMANI, AND P. A. WILSEY, *Analysis and Simulation of Mixed-Technology VLSI Systems*, Journal of Parallel and Distributed Computing (in press).

[19] A, BOUKERCHE, C, TROPPER, *Parallel Simulation on the Hypercube Multiprocessor*, Distributed Computing, Springer-Verlag, vol.8, no.4, pp 181-191.

THE DEVELOPMENT OF CONSERVATIVE SUPERSTEP PROTOCOLS FOR SHARED MEMORY MULTIPROCESSOR SYSTEMS

BOON-PING GAN†, YOKE-HEAN LOW†, WENTONG CAI‡, STEPHEN J. TURNER‡, SANJAY JAIN†, WEN JING HSU‡, AND SHELL YING HUANG‡

Abstract. This paper summarizes our work involving the successive refinements of a conservative superstep protocol. The refined protocols are known as the SST, SLS, and FET protocols. Each of the refinements improves the safetime bound, which in turn reduces the number of supersteps. In general, this helps to reduce the synchronization overhead and thus the execution time. However, a change in the number of supersteps can also introduce load imbalances within supersteps into the simulation. This observation becomes apparent with the FET protocol in particular. The performance of the refined protocols is compared through several semiconductor supply-chain simulation models. The results from the experiments strongly indicate the feasibility of using parallel discrete-event simulation techniques in the simulation of large scale systems such as a supply-chain.

1. Introduction. Parallel discrete-event simulation (PDES) has been a well-researched area for many years. An important area in PDES has been the design and development of new synchronization protocols. In PDES, a simulation is decomposed into logical processes (LPs) that can be simulated in parallel. PDES synchronization protocols are used to ensure that parallel execution of these LPs does not violate the causality constraint, i.e. events are strictly processed in timestamp order. PDES synchronization protocols can be broadly classified into two categories, conservative and optimistic. Conservative protocols [7, 4] strictly enforce the causality constraint by allowing the simulation to progress up to a safetime. A safetime is a time guarantee that no events with a lower timestamp will be received from the upstream LPs, and is normally computed based on events received from the upstream LPs. Unlike the conservative approach, optimistic protocols [8, 12] allow events to be executed without considering if executing those events will result in any causality violation. Each LP in this case will proceed to execute each incoming event as it arrives. If an event message arrives and has a timestamp that is lower than the ones that have been processed, the LP must correct this error by undoing those events that have been executed.

The focus of this paper is on the conservative protocol. Conservative protocols can be further sub-divided into two classes: synchronous [4] and asynchronous [7]. In the synchronous approach, LPs progress in supersteps. Each LP must wait for all the other LPs to complete their current superstep before the next superstep can proceed. In the asynchronous approach, each LP progresses in the simulation independently as long as there exist events that are

†Gintic Institute of Manufacturing Technology, 71 Nanyang Drive, Singapore 638075

‡Centre for Advanced Information Systems, School of Computer Engineering, Nanyang Technological University, Singapore 638798

safe to process. In this paper, we will provide a summary of our work in successively refining a synchronous conservative superstep protocol. A separate paper describing our work in using the asynchronous approach in a conservative synchronization protocol can be found in [9]. We will compare the performance of the different improvements made to the conservative superstep protocol using a semiconductor supply-chain simulation model.

The rest of the paper is organized as follows. Section 2 explains the original superstep protocol described in [4]. Section 3 describes the three refined superstep protocols that improve the safetime bound calculation. Section 4 discusses three important issues for the superstep protocols. In section 5, the semiconductor supply-chain simulation model used in the experiments will be explained. The performance of these protocols when running the supply-chain simulation model will also be presented and compared. Section 6 provides an overview of other related work carried out by researchers in the PDES community. Section 7 concludes this paper and outlines future research directions for our project. The proof of correctness of the four versions of the superstep protocol is given in the appendix.

2. Original superstep protocol. We will first describe the original conservative superstep protocol based on that presented in [4]. The conservative superstep protocol proceeds in a series of supersteps, where each superstep is followed by a barrier synchronization. Each superstep consists of two phases, the computation phase and the communication phase. This is analogous to the bulk synchronous parallel (BSP) model first proposed in [22]. The communication phase involves the exchange of event messages between the LPs. In the computation phase, each LP will compute a safetime for itself for the current superstep. The safetime is usually computed based on the events received from the upstream LPs. The LP can then simulate, in the current superstep, all events in its event-list with timestamp less than or equal to its safetime without violating any causality constraint.

The algorithm for the original conservative protocol is shown in Figure 2.1. In the algorithm, each communication link between two LPs is implemented by two buffers, which are priority queues, to take advantage of the shared memory architecture. The sender LPs will insert external events to one buffer while the receiver LPs will receive events from the other. The two buffers are swapped at the end of the superstep in the main simulation loop. With this approach, the conflict in accessing the same buffer between the sender and the receiver LPs is avoided. This eliminates the need for buffer locking.

The calculation of SafeTime (Part2(A) in Figure 2.1) uses both local and global information, and is set to be the maximum of LP_i.InClock and the global simulation time (GST). Intuitively, the LP_i.InClock provides the local information while GST constitutes the global information.

To calculate LP_i.InClock, a *link clock* is kept for each input link of LP_i (e.g. in the pseudo-code in Figure 2.1, LP_i.clock[k] holds the link clock value for the input link from LP_k to LP_i). It keeps track of the timestamp of the

```
// PART 1: Global initialization
Initialize the links between LPs and each LP's state.
GST = 0;
for all initial event e caused by InitialState do
  // assume event e scheduled for LP_i
  OrderInsert(e@e.TimeStamp, LP_i.event_q);
endfor

// PART 2: Executed by every LP, say LP_i.
while (GST < ∞) do

  // (A) calculate SafeTime
  LP_i.Inclock = ∞
  for each LP_k, s.t. LP_k has a directed link to LP_i do
    if(InBuff[LP_k][LP_i] ≠ empty) // update the link clocks
      LP_i.clock[k]=LastElementTime(InBuff[LP_k][LP_i])
      Merge InBuff[LP_k][LP_i] into LP_i's event-list, i.e. LP_i.event_q
    endif
    LP_i.InClock=min(LP_i.InClock, LP_i.clock[k]);
  endfor
  SafeTime = max(LP_i.InClock, GST);
  LP_i.out = ∞; // Tracks the smallest time of all external events sent out by
     LP_i

  // (B) simulate all safe events
  while (FirstElementTime(LP_i.event_q) ≤ SafeTime) do
    e = RemoveFirstElement(LP_i.event_q); // dequeue an event and process it
    LP_i.local_time=e.TimeStamp;
    LP_i.state = Simulate(e);
    for all InternalEvent ie caused by Simulate(e) do // enqueue new internal
     events
      OrderInsert(ie@ie.TimeStamp, LP_i.event_q);
    endfor
    for all ExternalEvent ee caused by Simulate(e) do //output external events
      // external event ee has timestamp equal to LP_i.local_time
      OrderInsert(ee@LP_i.local_time, OutBuff[LP_i][LP_j]) where ee is from LP_i
      to LP_j
      LP_i.out = min(LP_i.out, ee.TimeStamp);
    endfor
  endwhile

  // (C) calculate smallest timestamp of any event in this LP
         if (LP_i.event_q ≠ empty) then LP_i.MinTime =
     FirstElementTime(LP_i.event_q);
  else LP_i.MinTime = ∞; end if
  LP_i.MinTime = min(LP_i.out, LP_i.MinTime);

  // (D) global reduction to calculate new GST after all LPs reach
     barrier
  barrier_begin();
  GST = min_reduce(LP_i.MinTime);
  Swap InBuff and OutBuff
  barrier_end();

endwhile
```

FIG. 2.1. *Original Superstep Parallel Simulation Protocol*

last message received on the input link in the last superstep. The timestamp of the last message is readily available when the input link is implemented using priority queues ordered by the timestamp of events. LP_i.InClock is defined as the minimum value of all input link clocks. Thus, from the current superstep onwards, no event can arrive on an input link with a timestamp that is smaller than the LP_i.InClock value. As the calculation of LP_i.InClock assumes messages for communication links are received in non-decreasing timestamp order, all external events generated by an LP have timestamp equal to its local simulation time.

GST is defined as the minimum of the simulation time of all events waiting in the event-list, and the timestamp of all events that have just been generated in the current superstep. At the end of a superstep (Part 2(C) of the pseudo-code in Figure 2.1), each LP computes the minimum timestamp event that it knows about, by taking the minimum timestamp of events in its event-list, and those events that it generated in the current superstep. This minimum value is stored in the MinTime field. GST can then be computed by doing a global min-reduction of all the MinTime's. Therefore, GST is the smallest timestamp of any event in the whole system and so no event with a smaller timestamp than GST can be generated.

With the GST and LP_i.InClock values, SafeTime of the current superstep can then be computed. Each LP can proceed to simulate events with timestamp smaller than or equal to its SafeTime (Part 2(B) in Figure 2.1). The simulation terminates when GST reaches infinity (beginning of Part 2). This happens when there are no more events in the system. The algorithm is proved to be safe and deadlock free in the appendix.

3. Refinement of the original protocol. In this section, we discuss the three refined protocols which are improved versions of the original conservative superstep protocol. In each case, the improvement lies in the way the safetime of each LP is computed. In addition, the refined protocols all exploit lookahead information in the safetime computation. The original protocol does not use this lookahead information since external events are timestamped with the sender LPs' local simulation time. A technique called event pre-sending (to be discussed in Section 4) is used to further enhance the lookahead value for the three refined protocols. This has a significant impact on the performance of the protocols.

3.1. Sender-Simulation-Time protocol (SST). In this section, we will discuss the Sender-Simulation-Time (SST) protocol. The improvement to the safetime calculation is derived by noting that if there is a link from LP_k to LP_i, the original algorithm uses the link clock in the safetime calculation. However, we know that the timestamp of future events from LP_k must be greater than or equal to LP_k's local simulation time. Therefore, the local simulation time would give a more relaxed bound for LP_i's safetime calculation than the link clock used in the original protocol.

Figure 3.1 shows the algorithm for the modified conservative superstep protocol. The SST algorithm has essentially the same structure as the original algorithm. The two major differences are in Part 2(A), where the SafeTime is computed, and Part 2(B), where external events are pre-sent with timestamp equal to their occurence time.

In Part 2(A), for each LP_k that has a directed link to LP_i, LP_i computes the value

```
// PART 1: Global initialization
Initialize the links between LPs and each LP's state.
GST = 0;
for all initial event e caused by InitialState do
  // assume event e scheduled for LP_i
  OrderInsert(e@e.TimeStamp, LP_i.event_q);
endfor

// PART 2: Executed by every LP, say LP_i.
while (GST < ∞) do

  Merge events from all InBuff[LP_k][LP_i] for all LP_k connected to LP_i, into
     LP_i's event-list

  // (A) calculate SafeTime
  SafeTime = ∞
  for each LP_k, s.t. LP_k has a directed link to LP_i do
    SafeTime = min (SafeTime, LP_k.local_time + LookAhead[LP_k][LP_i])
  endfor
  SafeTime = max(GST, SafeTime)
  LP_i.out = ∞; // Tracks the smallest time of all external events sent out by
     LP_i

  // (B) simulate all safe events
  while (FirstElementTime(LP_i.event_q) ≤ SafeTime) do
    e = RemoveFirstElement(LP_i.event_q); // dequeue an event and process it
    LP_i.local_time=e.TimeStamp;
    LP_i.state = Simulate(e);
    for all InternalEvent ie caused by Simulate(e) do // enqueue new internal
      events
      OrderInsert(ie@ie.TimeStamp, LP_i.event_q);
    endfor
    for all ExternalEvent ee caused by Simulate(e) do //output external events
      Insert(ee@ee.TimeStamp, OutBuff[LP_i][LP_j]) where ee is from LP_i to LP_j
      LP_i.out = min(LP_i.out, ee.TimeStamp);
    endfor
  endwhile

  // (C) calculate smallest timestamp of any event in this LP
       if    (LP_i.event_q    ≠    empty)    then    LP_i.MinTime    =
     FirstElementTime(LP_i.event_q);
  else LP_i.MinTime = ∞; end if
  LP_i.MinTime = min(LP_i.out, LP_i.MinTime);

  // (D) global reduction to calculate new GST after all LPs reach
     barrier
  barrier_begin();
  GST = min_reduce(LP_i.MinTime);
  Swap InBuff and OutBuff
  barrier_end();
endwhile
```

FIG. 3.1. *Sender-Simulation-Time Protocol*

SafeTimeBound$_k$ = LP_k.local_time + LookAhead[LP_k][LP_i]. LP_k.local_time is the current simulation time of LP_k. LookAhead[LP_k][LP_i] represents the minimum advancement in simulation time required by LP_k to generate an external event for LP_i. The value SafeTimeBound$_k$ thus provides a time guarantee to LP_i that the next external event LP_k sends to LP_i will have timestamp no less than SafeTimeBound$_k$. LP_i then computes STB = the minimum of all its SafeTimeBound$_k$ values. The SafeTime of LP_i is taken to be the maximum of STB and GST.

In Part 2(B), each LP can simulate all events with timestamp smaller or equal to its SafeTime. Since the SafeTime calculation in this algorithm does not rely on the information on the link clocks, external messages need not be constrained to have their timestamps equal to the sender LP's local time. This allows external messages to be pre-sent with timestamp equal to their occurence time. The priority queue between LPs in the original algorithm can also be replaced by an unordered list and the OrderedInsert operation by a simple Insert. Any pre-sent external events received by an LP are held in the LP's event-list (ordered by timestamp of events) and are executed only when the LP's SafeTime becomes greater than or equal to the event's timestamp. The correctness of this protocol is proven in the appendix.

3.2. Sender-Last-SafeTime protocol (SLS). In this section, we describe the second improvement made to the conservative superstep protocol. We will refer to this algorithm as the Sender-Last-SafeTime (SLS) algorithm in subsequent sections.

We first note that at the end of each superstep n, each LP_k has simulated all events e that satisfy the condition $t_e \leq LP_k.S_n$, where t_e is the timestamp of event e and $LP_k.S_n$ is the safetime of LP_k for superstep n. The local time of LP_k is now set to the timestamp of the last event executed in this superstep. Since the safetime $LP_k.S_n$ for LP_k is a guarantee that any event it receives in the next superstep will have timestamp at least equal to or greater than $LP_k.S_n$, we can effectively take the safetime of LP_k as its local time at the end of a superstep and use this value to compute the safetime of its individual receiver LPs at the beginning of the next superstep.

Figure 3.2 shows the SLS algorithm. In this algorithm, we use a variable *LastSafeTime* to keep track of the safetime of each LP in the previous superstep. In Part 1, this variable is initialized to 0. Since each LP starts off with a local time of 0, we can use the individual lookahead values in the links to compute an initial safetime for each LP.

In the SLS algorithm, the only difference from the SST algorithm presented in section 3.1 is the computation of SafeTime in Part 2(A). For each LP_i, it computes a SafeTimeBound$_k$ value for each LP_k that has a directed link to it. However, instead of using the local time of LP_k, we now use the previous SafeTime of LP_k to compute SafeTimeBound$_k$. We set SafeTimeBound$_k$ to be the sum of LP_k.LastSafeTime and LookAhead$[LP_k][LP_i]$.

In Part 2(C), the GST computation is unchanged. However, in addition to the GST computation, each LP has to store the SafeTime value computed in the beginning of the current superstep in the LastSafeTime variable. The algorithm is proven to be correct in the appendix.

3.3. Future-Event-Time protocol (FET). The Future-Event-Time (FET) protocol is a further refinement of the SLS protocol, in which the safetime bound calculation is relaxed further. The FET protocol relaxes this bound by computing the safetime based on the sender LP's first safe event time in the current superstep instead of the sender LP's safetime at the last superstep. The

```
/* PART 1: Global initialization */
Initialize the links between LPs and each LP's state.
GST = 0;
for each LP_i
  LP_i.LastSafeTime=0.0;
endfor
for all initial event e caused by InitialState do
  // assume event e scheduled for LP_i
  OrderInsert(e@OccurrenceTime(e), LP_i.event_q);
endfor

// PART 2: Executed by every LP, say LP_i
while (GST < ∞) do

  Merge events from all InBuff[LP_k][LP_i] for all LP_k connected to LP_i, into
     LP_i's event-list

  // (A) calculate SafeTime
  SafeTime = ∞
  for each LP_k, s.t. LP_k has a directed link to LP_i
    SafeTime = min (SafeTime, LP_k.LastSafeTime + LookAhead[LP_k][LP_i])
  endfor
  SafeTime = max(GST, SafeTime)
  LP_i.out = ∞; // Tracks the smallest time of all external events sent out by
     LP_i

  // (B) simulate all safe events (same as SST)
                    .
                    .

  // (C) calculate smallest timestamp of any event in this LP (same as
     SST)
                    .
                    .

  LP_i.LastSafeTime = LP_i.SafeTime;

  // (D) global reduction to calculate new GST after all LPs reach
     barrier
  barrier_begin();
  GST = min_reduce(LP_i.MinTime);
  Swap InBuff and OutBuff
  barrier_end();

endwhile
```

FIG. 3.2. *Sender-Last-SafeTime protocol*

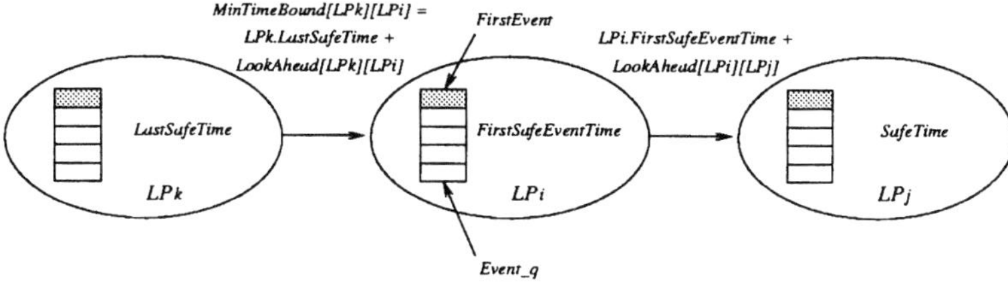

FIG. 3.3. *Computation of SafeTime in the FET Protocol*

first safe event time of any LP is defined as the timestamp of the first event in the LP that can be safely simulated in the current superstep (after merging of

```
/* PART 1: Global initialization */
Initialize the links between LPs and each LP's state.
GST = 0;
for each LP_i do
  LP_i.LastSafeTime=0.0;
  for each LP_k, s.t. LP_k has a directed link to LP_i do
    MinTimeBound[LP_k][LP_i] = LookAhead[LP_k] [LP_i]; endfor
endfor
for all initial event e caused by InitialState do
 OrderInsert(e@OccurrenceTime(e), LP_i.event_q); // assume event e scheduled
    for LP_i
endfor

// PART 2: Executed by every LP, say LP_i
while (GST < ∞) do

  Merge events from all InBuff[LP_k][LP_i] for all LP_k connected to LP_i, into
     LP_i's event-list

  // (A) calculate SafeTime
  // (A-1) compute LP_i.FirstSafeEventTime
  if (LP_i.event_q ≠ empty) then
    LP_i.FirstSafeEventTime=FirstElementTime(LP_i.event_q);
  else LP_i.FirstSafeEventTime=∞; endif
  for each LP_k, s.t. LP_k has a directed link to LP_i do
    LP_i.FirstSafeEventTime=min (LP_i.FirstSafeEventTime,
                                  MinTimeBound[LP_k][LP_i]);
  endfor
  barrier_synchronize();

  // (A-2) calculate LP_i.SafeTime
  L_i = ∞
  for each LP_k, s.t. LP_k has a directed link to LP_i do
    L_i=min ( L_i, LP_k.FirstSafeEventTime + LookAhead[LP_k][LP_i] ); endfor
  LP_i.SafeTime = max ( GST, L_i)
  LP_i.out = ∞; // Tracks the smallest time of all external events sent out by
     LP_i

  // (B) simulate all safe events  (same as SLS)
                          .
                          .

  // (C) calculate smallest timestamp of any event in this LP  (same
     as SLS)
                          .
                          .

  LP_i.LastSafeTime = LP_i.SafeTime;

  // (C-1) update MinTimeBound of receiver
  for each LP_j, s.t. LP_i has a directed link to LP_j do
    MinTimeBound [LP_i][LP_j] = LP_i.LastSafeTime+LookAhead[LP_i][LP_j];
  endfor

  // (D) global reduction to calculate new GST after all LPs reach
     barrier
  barrier_begin();
  GST = min_reduce(LP_i.MinTime);
  Swap InBuff and OutBuff
  barrier_end();

endwhile
```

FIG. 3.4. *Future-Event-Time Protocol*

external events to the local event list). By the definition of safetime, the first safe event time at the current superstep will definitely be greater than or equal to the previous superstep's safetime. Thus, the safetime bound imposed by the FET protocol is more relaxed. Figure 3.3 shows a graphical representation of the way safetime is computed using the FET protocol while Figure 3.4 shows the FET protocol algorithm.

Two new variables are introduced into the FET protocol, namely, MinTimeBound and FirstSafeEventTime, to realize the safetime computation described above. The MinTimeBound is used to determine the FirstSafeEventTime. In the beginning of each superstep, every LP will determine its own FirstSafeEventTime which will be used by the downstream LPs in the calculation of the safetime bound of the current superstep.

Referring to Figure 3.3, if MinTimeBound$[LP_k][LP_i]$ is the minimum time bound associated with the link from LP_k to LP_i, written as MinTimeBound$_k$, then any event arriving at LP_i from LP_k along the link must have timestamp greater than or equal to this bound. Therefore, MinTimeBound$_k$ is in fact the safetime bound that is imposed by LP_k to LP_i at the beginning of a superstep. Based on the SST protocol, this bound can be approximated using LP_k's previous superstep's SafeTime, LP_k.LastSafeTime, as follows:

$$\text{MinTimeBound}_k = LP_k.\text{LastSafeTime} + \text{LookAhead}[LP_k][LP_i]$$

Then, the MinTimeBound, MTB, can be determined as:

$$\text{MTB} = \min\{\text{MinTimeBound}_k | \forall LP_k \text{ that has a link to } LP_i\}$$

The MTB is then used to decide if the first event in LP_i is safe to be processed in the current superstep (i.e., to check if FirstElementTime(LP_i.event_q) $\leq$ MTB). If so, FirstSafeEventTime is equal to the timestamp of the first event in LP_i; otherwise, it is set to MTB. It is important to note here that the above must be done after all external events are merged into the LP's local event list. Otherwise, the event on top of the local event list might not be the one that has the smallest timestamp.

There is an extra barrier synchronization in this protocol, as it necessary to ensure that all the LPs have determined their FirstSafeEventTime before the safetime bound for LP_j can be calculated as follows:

$$\text{SafeTimeBound}_i = LP_i.\text{FirstSafeEventTime} + \text{Lookahead}[LP_i][LP_j]$$
$$\text{STB} = \min\{\text{SafeTimeBound}_i | \forall LP_i \text{ that has a link to } LP_j\}$$

As usual, SafeTime is set to the maximum of GST and STB. Note that LP_j's SafeTimeBound is determined by the FirstSafeEventTime of its upstream LPs' (e.g., LP_i), and the FirstSafeEventTime of LP_i is, in turn, determined by the LastSafeTime of its own upsteam LPs (e.g., LP_k).

The FET protocol in Figure 3.4 implements the SafeTime computation discussed above. In general, the structure of this protocol is the same as the SLS protocol. The two major differences are in part 2(A) for SafeTime computation and the introduction of part 2(C-1) into the FET protocol.

In part 2(A-1), each LP_i computes its own LP_i.FirstSafeEventTime. A new SafeTime is then computed after the barrier synchronization in part 2(A-2). In part 2(C-1), each LP_i computes the MinTimeBound that it imposes on its downstream LPs. This bound is used in the FirstSafeEventTime computation in the first section of part 2(A). The correctness of the protocol is proven in the appendix.

4. Discussions. In this section, the original superstep protocol and the three refined superstep protocols will be discussed based on three common issues. These issues are event pre-sending, simultaneous events, and zero lookahead.

4.1. Event pre-sending. Event pre-sending is a technique that is used by the three refined protocols to enhance their capability in exploiting the underlying model parallelism. To illustrate the idea of event pre-sending, let us consider a typical manufacturing scenario as follows: A machine M_k (modeled by LP_k) is processing a product arrival event e at time t_e. Assume that the product will proceed to machine M_i (modeled by LP_i) after this operation with service time of t_s on M_k and travelling time from M_k to M_i of t_d. A straight forward way to implement this scenario is to schedule a release event e' at time t_e+t_s on machine M_k. When the release event e' is simulated at time t_e+t_s, a corresponding arrival event e'' will be generated (with timestamp of $t_e+t_s+t_d$). It is obvious from this example that the lookahead between M_k (LP_k) and M_i (LP_i) is only t_d since the arrival event (e'') at M_i is only generated after the release event (e') at M_k is simulated.

Event pre-sending improves the lookahead value by generating the release event e' at M_k and the arrival event e'' at M_i together, with timestamp of t_e+t_s and $t_e+t_s+t_d$ respectively, when the product arrival event e is simulated in M_k. This guarantees a minimum lookahead of t_s+t_d between M_k (LP_k) and M_i (LP_i), which is greater than t_d as compared to the case where events are not pre-sent. In general, a larger lookahead value will translate to a better parallel performance.

Event pre-sending also offers another advantage. Consider the example described above using the SLS protocol but without event pre-sending. Assume that machine M_k (LP_k) is processing the product arrival event e at time t_e in superstep n with a safetime of $LP_k.S_n$, where $LP_k.S_n \geq t_e$. If $t_e+t_s > LP_k.S_n$, release event e' generated will only be executed in the next superstep $n+1$ and the arrival event e'' will only be visible to LP_i in superstep $n+2$.

Supposing the SLS protocol incorporates event pre-sending for the above example. Then the arrival event e'' will be visible to LP_i in superstep $n+1$. Assume LP_i has only one input link, which is from LP_k, then its safetime for superstep $n+1$, $LP_i.S_{n+1}$, would be equal to $LP_k.S_n+la$, where $la = t_s+t_d$. We assume here that GST for superstep $n+1$ is a smaller value than $LP_i.S_{n+1}$. Therefore, the arrival event e'' would have been executed in superstep $n+1$, since $e''.TimeStamp = t_e + la \leq LP_k.S_n + la = LP_i.S_{n+1}$.

Thus, event pre-sending allows both the release event e' and the arrival event e'' to be visible to LP_k and LP_i respectively at the same superstep. Both the events could be simulated in parallel if the safetime of the corresponding LPs permit.

It is not possible to apply event pre-sending on the original superstep protocol. This is due to the fact that events in the input buffer must be sent out in timestamp order. If event pre-sending is allowed, events may be sent out of timestamp order and the *InClock* computed will not be correct anymore.

4.2. Simultaneous events. Simultaneous events pose a problem for most conservative simulation protocols if the sequence in which the simultaneous events are resolved is important to the simulation model. They also introduce non-determinism into the simulation since multiple runs of the same simulation with the same inputs might not give the same simulation results. The superstep protocols previously discussed do not guarantee deterministic ordering in simulating simultaneous events. This is due to the fact that the use of safetime in the protocols has implicitly ordered the simultaneous events in certain ways depending on their order of arrival.

Let us consider an example. Assume the sequential simulation has two simultaneous events e_1 and e_2 both with timestamp t_e to be executed at LP_1, and that the model semantics requires the events be executed in the order $[e_1, e_2]$. The following situation may arise: LP_1 has safetime t_e, and event e_2 is at LP_1, but e_1 is still in transit to LP_1, and will only be received at the beginning of the next super-step. The protocol allows e_2 to be executed since the timestamp of e_2 is less than or equal to current safetime, and LP_1 will only simulate e_1 in the next super-step. The order $[e_1, e_2]$ has been violated.

This situation is unavoidable in the protocols described above because to enforce the order $[e_1, e_2]$ requires that LP_1 knows the state information of the LP that sends e_1. In general, to correctly implement the ordering function of the sequential simulation for a parallel context requires that LP_1 exchange events with other LP's to share their internal state information. Such state-information events are a source of overheads and degrade the performance of the parallel simulation. Because this problem is not unique to the super-step protocols, we assume that the model's semantics allows simultaneous events to be executed in any order. However, it may be possible to impose some ordering, e.g. within an LP in some situations, as will be discussed in the next subsection.

4.3. Zero lookahead. There is one major advantage of the superstep protocols as compared to some other conservative protocols – the protocols allow for zero lookahead. This is realized through the inclusion of GST in the safetime calculation. If GST is not included as part of the safetime calculation, it is possible for an LP's safetime to remain at a constant value. Therefore, GST is crucial in ensuring that the simulation progresses.

In the case when all the lookahead values are > 0, the safetime of all LP's strictly increases at each superstep for the SLS and FET protocols. This will eventually make it safe for the smallest-timestamp event to be simulated. The first implication for these protocols is that it is possible to remove GST from the safetime calculation without causing the simulation to deadlock. That is, the line:

$$LP_i.\text{SafeTime} = \max(\text{ GST}, LP_i.\text{SafeTime })$$

is eliminated [1]. This cannot be done on the original superstep protocol since it relies on GST to make safetime progress. If GST is removed from the above equation, the original superstep protocol will deadlock. Even though removing the GST calculation from the SLS and FET protocols is possible, one is not encouraged to remove this line since this computation allows an LP to push its safetime to GST within a superstep instead of going through multiple supersteps, in the case when GST is greater than the time bound imposed by the LP's upstream LPs.

The second implication of non-zero lookahead is that it is now possible to impose a certain ordering to the simultaneous events within an LP for the SLS and FET protocols. This has to do with the test condition in part 2(B) of Figure 3.1, 3.2, and 3.4:

$$(\text{ FirstElementTime}(LP_i.\text{event\_q}) \leq LP_i.\text{SafeTime })$$

The equality in the test condition is included to deal with the situation when $LP_i.\text{SafeTime} = \text{GST}$. It is needed to ensure that the events whose timestamp is equal to GST get simulated in the current superstep. But if all the lookaheads are non-zero, then the equality can be removed from the test condition. The test condition will then be:

$$(\text{ FirstElementTime}(LP_i.\text{event\_q}) < LP_i.\text{SafeTime })$$

By delaying the simulation of events time-stamped at safetime to the next superstep, it gives an LP the opportunity to receive all these events and perform some form of ordering before simulating them. This at least allows the simulation model to have the semantics for some restricted form of event-ordering within an LP. Furthermore, the deterministic ordering of events also offers the advantage of repeatable parallel simulation [2].

Removing the equality from the equation poses a problem to the original superstep protocol. This is due to its reliance on the equality to make progress in simulation time.

Based on the above discussion, it is obvious to note that it is impossible to achieve deterministic simulation if a protocol is designed to consider zero lookahead.

5. Experimental results. In this section, we will compare the performance of the different variations of the conservative superstep protocol using a supply-chain simulation model as a benchmark. Section 5.1 gives a brief description of the supply-chain simulation model. In section 5.2, the performance results of the SST, SLS and FET protocols will be discussed.

5.1. Supply-chain simulation model. The semiconductor supply-chain simulation model is constructed by linking up multiple wafer-fab models to a

[1] The other portions of the pseudocode remain unchanged because we use GST to terminate the protocol.

[2] When a simulation is said to be repeatable, it means that no matter how many times the simulation is run with the same inputs, the same output will be observed.

single assembly-and-test (A&T) model. The wafer-fab models are based on the Sematech data-sets [21]. Sematech is a consortium of semiconductor manufacturing companies that does research for its members in non-competitive areas of semiconductor manufacturing. Sematech has defined a data format to describe wafer fabrication processes, so that data can be easily exchanged among simulation users. There are six sets of factory data. Each factory data-set has six files, each describing an aspect of the factory environment.

The A&T model is a representative set of a logic and ASIC environment based on past studies in the integrated circuits (IC) industry. There are a total of six supply-chain models used in the experiments, models C1 to C6. To obtain the six supply-chain simulation models, we first start with six simulation models S1 to S6 each constructed using one of the six Sematech data-sets. We obtain the supply-chain simulation models C1 to C6 by linking a different number of simulation models S_i with the A&T model.

The matching of wafers to IC products was developed based on volume considerations; that is, large volume wafer products supply the large volume IC products. The transportation of wafers from wafer-fabs to the A&T facility is modeled as a short delay typical of the industry. The model features currently supported in the supply-chain simulator include wafer-lots batch, machine maintenance and failure, use of operator, setup rules and dispatching rules.

Further details on the supply-chain simulation model can be found in [11].

5.2. Results and performance analysis. The simulation runs are conducted on the semiconductor supply-chain simulation using the three refined protocols. No experimental results for the original superstep protocol will be presented here, due to the fact that the original superstep protocol performs extremely badly as compared to the improved protocols. The reason for its bad performance is because of its constraint in which events must be sent out at the LP's local simulation time. Thus, no lookahead information can be exploited with the original protocol.

The experiments are conducted on a 4 CPU (250MHz) Sun Ultra-Enterprise 3000 running Solaris 2.6. The compiler used is GCC 2.7.2. The parallel runtime library used in the experiments is Active-Threads [24], which provides a set of API similar to those normally found in other thread packages such as POSIX threads [10] and has been previously shown to be efficient [14]. We also modified a memory management package, VMalloc [23] to provide parallel memory allocation and deallocation.

In the simulation model, simulation objects (machines and operators) are mapped onto LPs, and the LPs are mapped onto processors. The mappings could either be 1-to-1 or many-to-1. We assume a 1-to-1 mapping of LPs to processors, and do not make any assumption on the simulation objects to LPs mapping. Thus, with only 4 CPU for experimentation, we group the simulation objects (machines and operators) to form 4 LPs. For this purpose, we used the partitioning package Metis [13] to partition the simulation objects onto the LPs. Past experiments [15] have shown partitions generated using an objective function that minimizes load-imbalance and maximizes lookahead values among LPs gives good performance.

Each simulation model has a simulation run length of one year. A first-come-first-served (FCFS) dispatching rule is used for lot scheduling on the machines. Table 5.1 shows the number of supersteps required for each variation of the conservative superstep protocol to complete each of the six supply-chain simulation models.

TABLE 5.1
Number of Supersteps Required by Different Variations of the Conservative Superstep Protocol for Supply-chain Simulation Models

	Number of Supersteps					
	C1	C2	C3	C4	C5	C6
SST Protocol	106539	51964	120285	30168	104823	14987
SLS Protocol	90949	47247	102190	29106	90949	14833
FET Protocol	81480	44820	90085	28388	81730	14684

TABLE 5.2
Execution Time of Supply-chain Simulation Models using Different Variations of the Conservative Superstep Protocol

	Execution Time (sec)					
	C1	C2	C3	C4	C5	C6
Sequential	247.77	410.34	245.22	203.76	305.99	500.43
SST Protocol	188.72	354.22	180.77	160.54	269.71	368.87
SLS Protocol	183.24	294.65	168.78	136.84	212.23	229.14
FET Protocol	184.14	284.32	165.83	138.42	227.88	272.26

TABLE 5.3
Speedup of Supply-chain Simulation Models using Different Variations of the Conservative Superstep Protocol

	Speedup					
	C1	C2	C3	C4	C5	C6
SST Protocol	1.31	1.16	1.36	1.27	1.13	1.32
SLS Protocol	1.35	1.39	1.45	1.49	1.44	2.18
FET Protocol	1.35	1.44	1.48	1.47	1.34	1.83

The number of supersteps required for the SLS protocol is smaller than that of the SST protocol across all six simulation models. The FET protocol took the least number of superstep to complete the six simulation models. This shows that the FET protocol is able to obtain a better safetime bound for each LP in the superstep computation as compared to the SST and SLS protocols.

However, the reduction in the number of supersteps does not necessarily guarantee a corresponding reduction in simulation execution time. Table 5.2

TABLE 5.4
Event time of Supply-chain Simulation Model C6 using Different Variations of the Conservative Superstep Protocol

	Event Granularity				
	LP_0	LP_1	LP_2	LP_3	*step_time*
SST Protocol	32.94	62.17	199.52	171.05	365.53
SLS Protocol	33.60	62.21	202.91	175.84	226.97
FET Protocol	37.18	64.87	208.47	177.93	270.71

shows the sequential execution time and the parallel execution time using different variations of the conservative superstep protocol. The corresponding speedup numbers are shown in Table 5.3. We note that the SLS protocol has consistently better execution time performance when compared to the SST protocol. However, the same cannot be said for the FET protocol. Although the FET protocol requires the least number of supersteps in all six simulation models, its execution time performance is worse than the SLS protocol in four out of six simulation models. Two possible factors could contribute to this degraded performance. One could be the higher synchronization overhead due to the extra synchronization barrier of the FET protocol. The other could be due to worse load imbalances within supersteps when the number of supersteps is reduced.

Supply chain simulation model C6 is profiled to identify the actual reason for the observation. The event granularity of each LP at each superstep during the simulation run is collected. The total event granularity of LP_i, $LP_i.etime$, and the total superstep execution time, *step_time*, is then computed using the equations shown below:

$$LP_i.etime = \sum_{n=1}^{s} LP_i.etime_n$$

$$step\_time = \sum_{n=1}^{s} max(LP_i.etime_n)$$

where $LP_i.etime_n$ is the event granularity of LP_i at superstep n and s is the total number of supersteps for the simulation run. The results are shown in Table 5.4, where simulation model C6 was executed with 4 LPs. The *step_time* is in fact the total execution time of the simulation run, excluding the protocol overhead. Thus, the protocol overhead can be computed by subtracting *step_time* of Table 5.4 from the execution time of Table 5.2 for the corresponding protocol. As can be seen, the differences of overhead among the protocols are marginal (less than 5 seconds). This observation rules out the first potential contributing factor for the worse performance of the FET protocol relative to the SLS protocol.

Referring back to Table 5.4, the total event granularity of the longest running LP (LP_2) is less than the *step_time*. This observation signifies that load imbalances within supersteps are happening during the simulation run. Apparently, the reduction in the number of supersteps with the refinement of the SLS protocol to the FET protocol makes the load imbalances worse. This suggests that being over aggressive in reducing the number of supersteps might not always be beneficial to the performance of parallel simulations. A similar argument was made in [5], where limiting the buffer size for events being sent by an LP at each superstep was proposed to contain the aggressiveness of the superstep protocol.

6. Related work. The superstep protocols presented in this paper belong to a class of conservative protocols known as the synchronous protocol. Some other synchronous protocols have also been proposed in the literature. These protocols differ in the way they compute safetime. Some of them make use of global information, while others make use of local information. Our protocols make use of both global (GST) and local (sender LP's simulation time, last-safetime, first-safe-event-time) information to compute the safetime of each LP.

In [1], a synchronous protocol based on a three phase algorithm (TPA) was proposed. It is a time window approach in which each LP computes its own safetime (known as upper bound of time window in the paper) by looking at all other LPs' minimum timestamped events in the system (global information). Lookahead information is used for the computation. Using this algorithm, every LP will have a different time window in a superstep. As its name suggests, this approach requires three barrier synchronizations per superstep.

The TPA algorithm was then improved in [2] to reduce its complexity in the safetime computation. An LP progressively computes its downstream LPs' safetime. The computation stops once the (source) LP's future event does not shrink the safetime of its successive downstream LPs. This approach helps to reduce the complexity of TPA by not looking at all LPs' minimum timestamped events. Therefore, this approach uses a limited amount of global information in the safetime computation. Comparing with our approach, the TPA and enhanced TPA require three barrier synchronizations while ours only requires one barrier (except for the FET algorithm which requires two). Also, the safetime computation of our approach does not need information from all the LPs in the system.

Lubachevsky also proposed a synchronous protocol in [19] whereby safetime is computed based on GST and events at LPs within the same region. This protocol is different from ours in two ways. Firstly, we do not limit an upper bound on LPs' safetime as long as it is safe for the LP to progress. Secondly, our protocol makes use of the information from immediate neighbouring LPs.

In [20], Nicol described a synchronous time-window algorithm that computes a safetime for all LPs based on a global minimum lookahead value. In constrast, our protocol allows the definition of a lookahead value for each directed link in the system and each LP computes its own safetime.

Chamberlain and Franklin [6] described synchronous conservative simulation using a global clock algorithm on a hypercube architecture. A performance prediction study was carried out on digital circuit simulations. The studies indicated a speedup of 9 on a 64-processor hypercube systems.

Ideas similar to those in the Future-Event-Time algorithm have been proposed for asynchronous conservative protocols in [3].

7. Conclusions and future work. In this paper, we have discussed our refinements to the conservative superstep protocol. These refinements have been successively incorporated in the SST, SLS and FET protocols respectively.

We have constructed supply-chain simulation models using real-world datasets and executed them using these different variations of the conservative superstep protocol. The experimental results indicate a consistent reduction in the number of supersteps as we further refine the conservative superstep protocol. But one important observation during the refinement process is that being over aggressive in computing safetime might not be beneficial to the performance of parallel simulation. This is especially true when the number of supersteps was reduced with the refinement from the SLS protocol to the FET protocol. Some form of constraining the aggressiveness of the superstep protocol is needed to ensure that load remains balanced within supersteps.

The performance results obtained from executing supply-chain simulation models using the different conservative superstep protocols strongly indicate the feasibility of developing applications that use this technology on a shared-memory platform. The next phase of the project will involve enhancing the modeling features of the supply-chain simulation, and studying the impact of these features on the parallel performance.

Acknowledgements. This research is supported by National Science and Technology Board, Singapore, under the project: *Parallel And Distributed Simulation for Virtual Factory Implementation.* The project is currently located at the Center for Advanced Information Systems at the Nanyang Technological University, Singapore. We would also like to acknowledge the contribution of Dr. Lim Chu Cheow whose work led to the writing of this paper.

REFERENCES

[1] A. Ayani, *Parallel Discrete-Event Simulation on Shared Memory Multiprocessors*, International Journal of Computer Simulation, Vol. 1 (1991), pp. 81–97.

[2] R. Ayani and H. Rajaei, *Parallel Simulation Based on Conservative Time Windows: A Performance Study*, Concurrency: Practice and Experience, Vol. 6, No. 2 (1994), pp. 119–142.

[3] A. Boukerche and C. Tropper, *SGTNE: Semi-Global Time of the Next Event Algorithm*, Proceedings of 9th Workshop on Parallel and Distributed Simulation, Lake Placid, USA (1995), pp. 68–77.

[4] W. Cai, E. Letertre and S. J. Turner, *Dag Consistent Parallel Simulation: a Predictable and Robust Conservative Algorithm*, Proceedings of 11th Workshop on Parallel and Distributed Simulation, Lockenhaus, Austria (1997), pp. 178–181.

[5] R. Calinescu *Conservative Discrete-Event Simulations on Bulk Synchronous Parallel Architectures*, Technical Report, Oxford University Computing Laboratory, PRG-TR-16-95 (1995).

[6] R. D. Chamberlain and M. A. Franklin, *Hierarchical Discrete-Event Simulation on Hypercube Architectures*, IEEE Micro, Vol. 10, No. 4 (1990), pp. 10–20.

[7] K. M. CHANDY AND J. MISRA, *Distributed Simulation: A Case Study in Design and Verification of Distributed Programs*, IEEE Trans. on Software Engineering, Vol. SE-5, No. 5 (1979), pp. 440–452.

[8] R.M. FUJIMOTO, *Optimistic Approaches to Parallel Discrete Event Simulation*, Transaction of the Society of Computer Simulation, Vol. 7 (1990), pp. 153–191.

[9] B.-P. GAN AND S. J. TURNER, *An Asynchronous Protocol for Virtual Factory Simulation on Shared Memory Multiprocessor Systems*, Journal of Operational Research Society, Special Issue on Progress in Simulation Research, Vol. 51, No. 4 (2000), pp. 413–422.

[10] INSTITUTE OF ELECTRICAL AND ELECTRONIC ENGINEERS, *Portable Operating System Interface (POSIX) - Part 1 Amendment 2: Threads Extensions [C Language]*, POSIX P1003.4a/D7 (1993).

[11] S. JAIN, C.-C. LIM, B.-P. GAN AND Y.-H. LOW, *Criticality of Detailed Modeling in Semiconductor Supply Chain Simulation*, 1999 Winter Simulation Conference, Phoenix, Arizona (1999), pp. 888-896.

[12] D. JEFFERSON AND H. SOWIZRAL, *Fast Concurrent Simulation Using The Time Warp Mechanism*, Distributed Simulation, La Jolla, California (1985), SCS-The Society for Computer Simulation, Simulation Councils, Inc, pp. 63–69.

[13] G. KARYPIS AND V. KUMAR, *A Fast and High Quality Multilevel Scheme for Partitioning Irregular Graphs*, SIAM J. on Scientific Computing, Vol. 20, No. 1 (1999), pp. 359-392.

[14] C.-C. LIM, Y.-H. LOW, W. CAI, W. J. HSU, S. Y. HUANG, S. J. TURNER, *An Empirical Comparison of Runtime Systems for Conservative Parallel Simulation*, 2nd Workshop on Runtime Systems for Parallel Programming, Orlando, Florida (1998). Also in Lecture Notes in Computer Science (No. 1388), Parallel and Distributed Processing, Jose Rolim (Ed.), Springer Verlag, pp. 123-134.

[15] C.-C. LIM, Y.-H. LOW, B.-P. GAN, S. J. TURNER, S. JAIN, W. CAI, W. J. HSU AND S. Y. HUANG, *A Parallel Discrete-Event Simulation of Wafer Fabrication Processes*, 3rd High Performance Computing (HPC) Asia, Singapore (1998), pp. 1180–1189.

[16] C.-C. LIM, Y.-H. LOW AND S. J. TURNER, *Relaxing Safetime Computation of a Conservative Simulation Algorithm*, Proceedings of 1998 International Conference on Parallel and Distributed Processing Techniques and Applications, Las Vega, Nevada (1998), pp. 1538–1545.

[17] C.-C. LIM, Y.-H. LOW AND B.-P. GAN, *Computing Safetime in a Conservative Synchronous Simulation Based on Future Events*, International Conference on Parallel and Distributed Processing Techniques and Applications, Las Vegas, Nevada (1999).

[18] Y.-H. LOW, C.-C. LIM, B.-P. GAN, S. JAIN, W. CAI, W. J. HSU, S. Y. HUANG AND S. J. TURNER, *Conservative Parallel Simulation for Manufacturing System*, 8th International Parallel Computing Workshop, Singapore (1998), pp. 293–300.

[19] B. D. LUBACHEVSKY, *Efficient Distributed Event-Driven Simulations of Multiple-Loop Networks*, Communications of the ACM, Vol. 32, No. 1 (1989), pp. 111–131.

[20] D. NICOL, *The Cost of Conservative Synchronization in Parallel Discrete Event Simulation*, Journal of the ACM, Vol. 40, No. 2 (1993), pp. 304–333.

[21] SEMATECH, *Modeling Data Standards, version 1.0*, Technical report, Sematech, Inc., Austin, TX78741, 1997.

[22] L.G. VALIANT, *A Bridging Model for Parallel Computation*, Communication of the ACM, Vol. 33, No. 8 (1990), pp. 103–111.

[23] K.-P. VO, *Vmalloc: A General and Efficient Memory Allocator*, Software - Practice and Experience, Vol. 26, No. 3 (1996), pp. 357–374.

[24] B. WEISSMAN, B. GOMES, *Active Threads: Enabling Fine-Grained Parallelism*, Proceedings of 1998 International Conference on Parallel and Distributed Processing Techniques and Applications, Las Vegas, Nevada (1998), pp. 115–122.

Appendix A. Correctness. In this section, the correctness of the four superstep protocols will be proved. Basically, the proof covers three aspects of the protocols. Firstly, the protocol is proved to be safe by proving that future events received by an LP have a timestamp greater than or equal to the current superstep safetime. Next, the protocol is proved to be deadlock free. Lastly, we prove that the protocol terminates correctly. The proof on deadlock and termination is common to all four protocols. Thus, it will only be presented once in the proof of the original superstep protocol.

A.1. Original superstep protocol.

Safety. The calculation of safetime for the original superstep protocol involves two components:

1. Calculation of input link clocks:
   The events in each buffer are in timestamp order, no future event can arrive on an input link with a timestamp which is smaller than the link clock value.
2. Calculation of GST_n:
   GST_n is the minimum timestamp of all un-simulated events in the system at the beginning superstep n. It is therefore the smallest timestamp of any event in the whole system and so no event can be generated with a smaller timestamp then GST_n.

Let $LP_i.InClock_n$ be the minimum of all the input link clocks for LP_i in superstep n. The safetime of LP_i in superstep n, $LP_i.S_n$, is then calculated as the *maximum* of $LP_i.Inclock_n$ and GST_n. Thus, there are two cases to consider:

a) $LP_i.Inclock_n \geq GST_n$
   This will be the normal situation since each input link clock is updated using the events in the input buffer for that link. These events are still to be processed and so they will all have a timestamp which is greater than or equal to GST_n.
   It is safe to set $LP_i.S_n = LP_i.Inclock_n$, since we know that no future event can arrive on any of the LP's input links with a timestamp which is smaller than $LP_i.Inclock_n$.
b) $LP_i.Inclock_n < GST_n$
   This situation can arise when one or more of the input buffers of the LP are empty, so that the corresponding link clock is not updated. It is the absence of messages on an input link which leads to the possibility of deadlock in conservative approaches. Feed-back loops in the simulation application are likely to result in the situation where $LP_i.Inclock_n < GST_n$.
   It is safe to set $LP_i.S_n = GST_n$, since we know that in the whole system, no event can be generated with a timestamp which is smaller than GST_n.

From the above, we can conclude that our algorithm guarantees that there is no violation of causality since timestamps of future external events are greater than or equal to the safetime computed.

Deadlock and termination. Next, we will show that deadlock cannot occur with the original superstep protocol We note that for each LP_i, we have $LP_i.S_n \geq GST_n$ whichever of the above cases holds. Since GST_n is the smallest timestamp of any event in the whole system, at least those events with a timestamp equal to GST_n will be executed in the current cycle. In each superstep of the simulation, there must be at least one event with a timestamp equal to GST_n, so the simulation will progress and deadlock will not occur.

Lastly, we prove that the superstep protocol terminates correctly when $GST_n = \infty$. This means that each LP finds no event in its internal event-list or its external output buffers, which implies that there are no events left in the entire system.

A.2. Sender-Simulation-Time protocol. For the Sender-Simulation-Time (SST) protocol, the safetime of an LP is computed based on the sender LPs' local simulation time at the end of the superstep. As discussed at the beginning of the appendix, we will prove that future external events arriving at an LP will have timestamps greater than or equal to the LP's safetime. This is done by using a simple induction method.

Induction Argument. For any LP_i, if

$$\forall e \in LP_i, e.TimeStamp \geq LP_i.S_{n-1}$$

holds at the beginning of superstep n, then

$$\forall e \in LP_i, e.TimeStamp \geq LP_i.S_n$$

is true at the beginning of superstep $n+1$.

Now, we prove that the induction argument is correct. Consider any event, e', simulated by LP_k in superstep n. By the induction assumption, event e' must satisfy the following condition:

$$LP_k.S_{n-1} \leq e'.TimeStamp \tag{A.1}$$

If event e' generates any external event e scheduled for LP_i, by the definition of lookahead, the following condition must be satisfied:

$$e.TimeStamp \geq e'.Timestamp + LookAhead[LP_k][LP_i]$$

From inequality (A.1), we then have

$$e.TimeStamp \geq LP_k.S_{n-1} + LookAhead[LP_k][LP_i] \tag{A.2}$$

The safetime of LP_i at superstep n, $LP_i.S_n$, is taken to be the maximun of GST_n and STB_n. GST_n is the minimum timestamp of all un-simulated events in the system at the beginning of superstep n, and STB_n is calculated as follows:

$$STB_n = \min\{LP_l.LocalTime_n + LookAhead[LP_l][LP_i] \mid \forall LP_l \text{ with a link to } LP_i\} \tag{A.3}$$

where $LP_l.LocalTime_n$ is the local simulation time of LP_l at the beginning of superstep n. Therefore, there are two cases to consider here:

a) $LP_i.S_n = GST_n$
Since GST_n is the minimum timestamp of all un-simulated events at the beginning of superstep n, this derives the inequality of $e.TimeStamp \geq GST_n = LP_i.S_n$. So, the induction argument is correct.

b) $LP_i.S_n = STB_n$
In this case, we have (from equation (A.3))

$$LP_i.S_n = STB_n \leq LP_k.LocalTime_n + LookAhead[LP_k][LP_i]$$

Since the timestamp of any event simulated in superstep $n-1$ is less than or equal to the safetime at superstep $n-1$, $LP_k.LocalTime_n \leq LP_k.S_{n-1}$. Then, we have

$$LP_i.S_n \leq LP_k.S_{n-1} + LookAhead[LP_k][LP_i]$$

This brings us to $e.TimeStamp \geq LP_i.S_n$ based on inequality (A.2), which also signifies that the induction argument is correct.

If we regard program initialization as the very first superstep (refer to Figure 3.1), the condition

$$\forall e \in LP_i, e.TimeStamp \geq LP_i.S_0$$

is obviously true since $e.TimeStamp \geq 0$ and $LP_i.S_0 = 0$.

Based on the above discussion, we can conclude that LP_i will not receive any event strictly earlier than its current safetime. The safetime calculation in SST protocol is safe.

A.3. Sender-Last-SafeTime protocol. For the Sender-Last-Safetime (SLS) protocol, the safetime of an LP is computed based on its upstream LPs' previous superstep safetimes. The correctness of the safetime computation is proven by referring to the proof of the SST protocol discussed in the last section.

Similar to the SST protocol, in the SLS protocol the safetime of LP_i at superstep n is also set to the maximum of GST_n and STB_n. However, STB_n is calculated as follows:

$$STB_n = \min\{LP_l.S_{n-1} + LookAhead[LP_l][LP_i] \mid \forall LP_l \text{ with a link to } LP_i\} \quad \text{(A.4)}$$

We need to prove the induction argument given in Section A.2 is also correct for the SLS protocol. Suppose event e' simulated by LP_k at superstep n generates an external event e for LP_i. For case a), where $LP_i.S_n = GST_n$, the proof is the same as the one for the SST protocol. For case b), where $LP_i.S_n = STB_n$, from equation (A.4) we know that

$$LP_k.S_n = STB_n \leq LP_k.S_{n-1} + LookAhead[LP_k][LP_i]$$

Therefore, from inequality (A.2), we will have $e.TimeStamp \geq LP_i.S_n$, which signifies that external events generated for LP_i in superstep n have timestamps larger than or equal to LP_i's safetime of superstep n. With this, we can conclude that the saftetime calculation in the SLS protocol is safe.

A.4. Future-Event-Time protocol. For the Future-Event-Time (FET) protocol, the safetime of an LP is computed based on its upstream LPs' first safe event times. To prove the correctness of the safetime computation for the FET protocol, we make use of the proven fact of the SLS protocol.

The idea of the FET protocol is to substitute $LP_l.S_{n-1}$ by $LP_l.FirstSafeEventTime$ in computing the safetime bound (see equation (A.4)), where
$LP_l.FirstSafeEventTime \geq LP_l.S_{n-1}$. From the induction argument of the preevious section, external events generated for LP_l at superstep $n-1$ will have timestamp $\geq LP_l.S_{n-1}$. If we can guarantee that there is no event occurring in the interval $[LP_l.S_{n-1}, LP_l.FirstSafeEventTime)$, then $LP_l.FirstSafeEventTime$ would be a correct and better substitute for the $LP_l.S_{n-1}$ in the safetime computation. Note that external events generated at superstep $n-1$ are merged into the local event list at the beginning of each superstep. Thus, we only need to prove that if e is the future event arriving at LP_l (starting from superstep n)

$$e.TimeStamp \geq LP_l.FirstSafeEventTime$$

$LP_l.FirstSafeEventTime$ is essentially a lower bound on the timestamp of the first event that LP_l will safely execute after superstep $n-1$. This first safe event is either already in LP_l's local event list or will be arriving from one of the upstream LPs of LP_l. From Figure 3.4, the first safe event time of LP_l, $LP_l.FirstSafeEventTime$, is computed as follows:

$$LP_l.FirstSafeEventTime = \min(MTB, FirstElementTime(LP_l.event\_q))$$

where, MTB is defined as:

$$\text{(A.5)}\, MTB = \min\{MinTimeBound[LP_p][LP_l] \mid \forall LP_p \text{ with a link to } LP_l\}$$

and

$$MinTimeBound[LP_p][LP_l] = LP_p.S_{n-1} + LookAhead[LP_p][LP_l]$$

Therefore, there are two cases to consider here:

a) $LP_l.FirstSafeEventTime = MTB$

Assume that LP_q generates an event e for LP_l after superstep $n-1$. From equation (A.5), we know that at superstep n,

$$MTB \leq MinTimeBound[LP_q][LP_l] = LP_q.S_{n-1} + LookAhead[LP_q][LP_l]$$

According to inequality (A.2) the timestamp of event e must satisfy the following:

$$e.TimeStamp \geq LP_q.S_{n-1} + LookAhead[LP_q][LP_l]$$

So, we have

$$e.TimeStamp \geq MTB = LP_l.FirstSafeEventTime$$

In this case $LP_l.FirstSafeEventTime \leq FirstElementTime(LP_l.event\_q)$. So, there would not be any event in the event queue that has a timestamp less than $LP_l.FirstSafeEventTime$. Therefore, there is no event that occurs at LP_l in the interval $[LP_l.S_{n-1}, LP_l.FirstSafeEventTime)$.

b) $LP_l.FirstSafeEventTime = FirstElementTime(LP_l.event\_q)$
Similarly, we assume that LP_q generates an event e for LP_l after superstep $n-1$.
We know that

$$FirstElementTime(LP_l.event\_q) \leq MTB$$
$$MTB \leq MinTimeBound[LP_q][LP_l], \text{ and}$$
$$MinTimeBound[LP_q][LP_l] = LP_q.S_{n-1} + LookAhead[LP_q][LP_l]$$

Thus, based on inequality (A.2), we have

$$e.TimeStamp \geq FirstElementTime(LP_l.event\_q) = LP_l.FirstSafeEventTime$$

This shows that in the interval $[LP_l.S_{n-1}, LP_l.FirstSafeEventTime)$, no events will occur at LP_l.

Therefore, we can conclude that the safetime calculation in the FET protocol is also safe.

IMPLEMENTATION OF A VIRTUAL TIME SYNCHRONIZER FOR DISTRIBUTED DATABASES ON A CLUSTER OF WORKSTATIONS *

A. BOUKERCHE†, S. K. DAS‡, A. DATTA§, AND T. E. LEMASTER¶

Abstract. The availability of high speed networks and improved microprocessor performance has made it possible to build inexpensive cluster (or network) of workstations as an appealing vehicle for parallel and distributed computing. In this paper, we study the performance of a distributed database system synchronized, by *virtual time* (VT) mechanism by experimenting on a LAN connected collection of Sun SPARC workstations. To the best of our knowledge, ours is the first attempt to report the virtual time performance in distributed databases using a cluster of workstations. The experimental results demonstrate that the VT synchronization approach is a viable concurrency control method. Compared to a widely used multiversion (MV) concurrency control scheme based on timestamps order, the VT synchronization scheme is shown to yield about 25-30% reduction in the response time. The reason for choosing multiversion concurrency control as the basis for comparison is that both VT and MV protocols are based upon the concept of timestamps, and attain high performance by using multiple version of data objects. (The term "version" refers to the value of data item produced by a transaction that's either active or committed.)

Key words. Distributed Database, Virtual Time, Optimistic Method, Cluster of Workstations

1. Introduction. In recent years, the availability of high speed networks and improved microprocessor performance have made it possible to build inexpensive cluster (or network) of workstations as an appealing vehicle for parallel and distributed computing. In this paper, we focus on distributed database concurrency control mechanisms. Although distributed databases have been an area of extensive research, to the best of our knowledge, there are no reported results on the performance of virtual time synchronization in parallel and distributed databases using a cluster of workstations. The goal of this paper is to implement a distributed database system, synchronized by the *virtual time* (VT) mechanism, using a LAN connected collection of Sun SPARC workstations; and study its performance in comparison with a widely used concurrency control protocol, namely the multiversion (MV) scheme based on timestamps order.

Virtual Time is a fundamental concept in optimistic synchronization schemes and many applications of virtual time have been proposed, including parallel and distributed simulations. The Time Warp (TW) [15, 16, 19] based mechanism is a technique to implement virtual time by allowing application programs to cancel previously scheduled events. While conservative synchronization methods

*Part of this work was supported by UNT Faculty Research Grant and Texas Advanced Technology Program Grant ARP/TATP.

†Dept. of Computer Science, University of North Texas (Contact Person: boukerche@cs.unt.edu)

‡Dept. of Computer Science& Engineering, University of Texas at Arlington

§Dept. of Computer Science, University of Nevada at Las Vegas

¶Dept. of Computer Science, University of Nevada at Las Vegas

[16, 9] rely on blocking to avoid violation of dependency constraints, Time Warp relies on detecting synchronization errors at run time and recovering them by rollback mechanisms. Good surveys of these protocols can be found in [16]. The main advantages of Time Warp over more conventional conservative protocols are that the former offers the potential for greater exploitation of concurrency and parallelism and, more importantly, greater transparency of the synchronization mechanism to the programmer. The transparency is due to the fact that Time Warp is less reliant on application specific information such as which computations depend on one another.

Despite the fact that active research on virtual time has been pursued in the past decade, the actual performance study of *distributed database systems* (DDBS) synchronized by virtual time has not gained much attention. Jefferson [18, ?] suggested the use of virtual time to synchronize DDBS, and some related theoretical work was reported in [24]. Therefore, a pragmatic question is whether virtual time is really a viable concurrency control method for distributed database systems and whether it can provide good performance when implemented on a distributed computing platform.

In this paper, we investigate the efficiency of a DDBS synchronized by virtual time (VT) by implementing it and analyzing its performance on a cluster of workstations. We implement an optimistic synchronization protocol by making use of a Time Warp based mechanism[1]. We also compare the performance of our method with a MultiVersion Timestamp Ordering (MV) concurrency control method [3, 13, 14, 21]. Empirical results demonstrate that the VT approach outperforms the MV protocol and about 25-30% reduction in the response time is observed. A preliminary version of this paper appeared in [6].

We choose to compare the VT synchronization scheme with the MV protocol because they are conceptually similar in the sense that the notion of "pseudotime" in MV is similar to that of virtual time, in addition to the fact that MV is a widely used technique in DDBS. Furthermore, both of these methods have the following advantages: (1) they exhibit high concurrency level; (2) there is no need for deadlock detection and recovery; and (3) they are highly efficient unless there are conflicts, which are resolved by undoing operations or aborting transactions.

The remainder of this paper is organized as follows. Section 2 reviews the basic components of a DDBS while Section 3 summarizes previous and related work on concurrency control methods used in a DDBS. In Section 4, we describe virtual time as a concurrency control mechanism for distributed databases. It also describes the data structures, rollback mechanism, and control algorithms used in our implementation. Section 5 outlines the MV concurrency control method and Section 6 presents the details of our performance study. A comparison between VT synchronization in DDBS and the MV method is also presented. The last section offers conclusions.

[1]In the sequel, *virtual time* will refer to the time warp based mechanism.

2. Distributed Database (DDBS) Model . We view a distributed database system as a collection of logically interrelated databases distributed amongst a group of computer nodes. We assume that two processes can exchange messages whether they are located at the same site or at different nodes (in which case the messages are sent over the communication network).

Each node is a centralized database system (DBS), which stores a portion of the database. *Transactions* are the basic units of work in database systems, and they may involve the generation of a number of sub-transactions. Basically, a transaction is the execution of a sequence of *read* and *write* operations on database, and (transaction) operations such as commit and abort. A *commit* guarantees the permanence of the effects of a transaction, while an *abort* removes the partial effects of a transaction. We assume that each transaction is self-contained, meaning that it performs its computation without any direct communication with other transactions. Transactions do communicate indirectly, of course, by storing and retrieving data in the database. However, this is the only way they can affect each other's execution.

In the DDBS model, each transaction consists of one or more processes (subtransactions) that execute at one or more nodes. To ensure that transactions do not violate certain concurrency control problems, protocols are used to constraint the sequencing of database operations submitted by transactions.

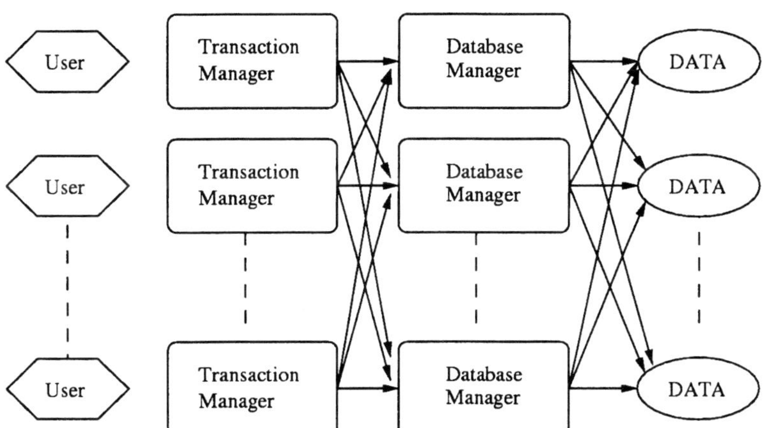

FIG. 2.1. **Structure of a Database System**

As illustrated in Figure 2.1. the general structure of the DDBS model consists of four major components at each node: a *transaction*, a *transaction manager* (TM), a *database manager* (DM), and a *data*. Transactions are generated from users. The TM supervises the interaction between users and the DDBS, while the DM controls access to the data and provides an interface between the user (via the query) and the file system. It is also responsible for controlling the simultaneous use of the database and maintaining its integrity and security.

Transactions submit their operations to the TM, which passes them to the scheduler. The scheduler receives *read*, *write*, *commit* or *abort* operations from TM. The scheduler can pass aborts to the DM immediately. For read, write, or commit operations, the scheduler must decide, possibly after some delay, whether to reject or accept the operation. If it rejects the operation, the scheduler sends a negative answer acknowledgement to the TM, which sends an abort back to the scheduler, which in turn promptly passes the abort to the DM. If the scheduler accepts the operation, it sends it to the DM, which processes it by manipulating the database. When the DM has finished processing the operation, it acknowledges to the scheduler, which passes the acknowledgement to the TM. For a read, the acknowledgement includes the value read.

When a TM receives a transaction's read or write that cannot be serviced at its site, the TM forward that operation to the DM at another site that has the data needed to process the operation. Thus, each TM can communicate with every DM by sending messages over the network.

3. Concurrency Control and Recovery. When two or more transactions execute concurrently, their database operations execute in an *interleaved* fashion. That is operations from one program may execute in between two operations from another program. This interleaving can cause programs to behave incorrectly, thereby leading to an inconsistent database. The main goal of concurrency control is to avoid these types of errors by controlling the execution order of database operations. This execution order is called a *schedule*, and it is, in general, correct if it is equivalent to any other schedule in which transactions are executed serially, i.e., the schedule is said to be *serializable.* Protocols that enforce serializability are called *concurrency control protocols.* Closely related are *recovery* protocols. While serializablity guarantees the consistency of data, *recoverability* guarantees that after a failure condition (transaction failure, system crash, etc.), the database can be recovered to a consistent state at which point things can carry out as usual. Recoverability is maintained by not allowing a transaction to commit until all of its data has been committed.

3.1. Previous and Related Work. A number of concurrency control schemes have been proposed in the literature to ensure non-interference or isolation of concurrently executing transactions. Accordingly, several protocols and algorithms have also been proposed to ensure serializability of schedules [13, 14] One important class of algorithms employs the technique of locking data items to prevent multiple transactions from accessing the items concurrently [14, 20]. The locking mechanism, by enforcing *two-phase locking* (2PL) rules, also guarantees serializability [14]. A basic 2PL scheduler manages and uses its lock according to the following three rules: Rule 1 prevents two transactions from concurrently accessing a data in conflicting modes. Thus, conflicting operations are scheduled in the same order in which the corresponding locks are obtained. Rule 2 supplements rule 1 by ensuring that the DM processes operation on a data in the order that the scheduler submits them. Rule 3 is called two-phase rule (hence the name name two-phase locking). Each transaction might be divided into two phases: *growing phase* during which it obtains locks, and a shrinking phase during which it releases locks. The basic idea of rule 3 is to guarantee that all pairs of conflicting operations of two transactions are scheduled in the same manner. However, the use of locks can cause two additional problems: *deadlock* and *livelock*. Moreover, the 2PL locking may limit the amount of concurrency that can occur in a schedule. Several variations of the 2PL have been reported in the literature. Interested readers may consult [3, 14] for further details.

Another class of concurrency control algorithms uses transaction timestamps. A *timestamp* is a unique identifier created by the distributed database system (DDBS) to identify a transaction, and it is typically assigned in the order in which the transaction is submitted to the system [3, 14, 22]. Therefore, a timestamp can be thought of as a transaction start time. Concurrency control algorithms based on timestamps do not use locks; hence deadlocks cannot occur.

The timestamp ordering protocol, like two-phase locking, also guarantees serializability of schedules. Although deadlock does not occur with timestamp ordering, cyclic restart (and hence *starvation*) may occur if a transaction is continually aborted and restarted. For details on timestamps based concurrency control techniques, refer to [3, 13, 14].

Multiversion Timestamp Ordering (MV) protocols have been proposed in [3, 23]. The basic idea here is to keep several versions (values) of an item. When a transaction needs to read data item for which multiple versions exist, the DDBS selects one of the versions. The value read by a transaction must be consistent with some serial execution of the transaction with single version of the database. Thus, the concurrency control problem is transferred into the selection of the correct version from multiple version of a data item. With multiversion technique, write operations can occur concurrently, since they do not overwrite each other. Furthermore, the read operation can read any version. This results in greater flexibility in scheduling concurrent transactions. Many schemes have been proposed for controlling concurrency using the MV approach [13, 14]. Other recent work on concurrency control includes semantic-based technique [2], transaction models for long running activities [11], and multilevel transaction management [22].

In [5], the authors presented a concurrency control and recovery protocol for distributed databases that produces a schedule that is equivalent to that of a temporally ordered serial schedule. The algorithm, which they refer to as T3C, allows transactions to be received out of order within a window of tolerance. It doesn't required declaration of readsets, while the writesets are declared simply at the table level and without predicates. The T3C algorithm is basically desirable in the case where both in-order transaction execution and semantic correcteness are important.

3.2. Parallel and Distributed Simulation. Interestingly enough, there are lot of similarities between concurrency control protocols and parallel/distributed simulation protocols where causal events must also be executed before their resultant events. Synchronization protocols for parallel and distributed simulation can be classified into two groups: *conservative* and *optimistic*. While conservative synchronization techniques [9, 16] rely on *blocking* to avoid violation of dependence constraints, *optimistic* methods [19] rely on detecting synchronization errors at run-time and on recovery using a *rollback* mechanism and a virtual time paradigm. (For a good survey, refer to [4].)

As opposed to conservative synchronization protocols, we believe that virtual time using a time warp scheme for synchronization is an elegant concept in DDBS [18]. The virtual time synchronization algorithm has the following advantages over conservative approaches to database systems: (1) Time Warp has been shown to be deadlock free; (2) entire transactions are not aborted and/or restarted since individual actions within a transaction are rolled back when conflict occurs; (3) it is free from such requirements as predeclaration, sequentiality, or two-phase structure; and (4) it adheres to an object oriented approach to database design in which there are no formal distinctions among *transactions*, *data*, and *system* objects.

The similarity between protocols for concurrency control and parallel/distributed simulations can be summarized as follows. While the multiversion (MV) timestamp ordering protocol has been found to yield good performance enforcing serializability in highly concurrent systems, virtual time (VT) synchronization based upon Time Warp paradigm has been found to be very successful in optimistic parallel and distributed simulations.

4. Virtual Time Synchronization Protocol. Timestamps in DDBS can be generated in many ways. One possibility is to use a counter that is incremented each time its value is assigned to a transaction. Another way to implement timestamps is to use the current value of the local clock and ensure that no two timestamp values are generated during the same tick of the clock. These local timestamp-generating schemes can be kept fairly well synchronized by including timestamp in the messages sent between sites. A virtual time (VT) synchronization scheme might be used to generate timestamps and as a concurrency control mechanism.

4.1. Design Requirements. Virtual time can be viewed as a global, one-dimensional temporal coordinate system imposed on a distributed system to measure the computational progress and define synchronization [19]. There are two important properties of virtual time. First, it may increase or decrease with respect to real time and there is no restriction on the amount or frequency by which it changes. Second, the virtual times generated must form a total (or partial) ordering on the relation "less than."

In our model, DDBS is viewed as a collection of of data and transaction[2] objects (sequential processes) that execute concurrently and communicate through

[2] A transaction may involve the generation of a number of sub-transactions, each of which might be executed at a different site. Hence, a transaction may consist of a set of objects in our DDBS model.

timestamped messages. Each object is capable of both computation and communication. In the VT synchronization model, a *local virtual time* (LVT) is associated with each object. The LVT identifies how far the node has been advancing its state forward. There are no formal distinctions between the transaction and data objects. While data objects respond only to *read/write* messages, transaction objects receive messages directly from external users. Objects communicate with each other through *Request* and *Response* primitives which send messages. A *Request* either reads or writes a data value, while a *Response* returns a data value. After an object issues a *Request*, it blocks to await a response. The possibility of deadlock exists since two objects within a transaction could request data values from each other at the same time, and each object will be waiting for the other to respond to the *Request*. As we shall see in the next section, the VT scheme prevents deadlock between transactions since all transactions are executed in timestamp order and all such timestamp are unique.

An object (or process) may send a message to any other object at any time, without any restrictions placed upon potential communication paths between processes. No fixed or static declarations are required before execution begins. The communication medium is assumed to be reliable but messages are not required to arrive in the order in which they are sent. Indeed, different transactions occasionally cause an unordered message to be received by an object. When this occurs, *rollback* is used for (messages) synchronization. Since a transaction may consist of a set of objects (i.e., sub-transactions), instead of rolling back the entire transaction, only certain actions (such as those that are causally connected to the errant message) within the transaction are rolled back. This is where the virtual time synchronization approach diverges from a (traditional) conservative protocol, in which the entire transaction is either aborted and retried or is blocked. Recall that a transaction may involve the generation of a number of sub-transactions, each of which might be executed at a different site. The algorithms include transaction objects (TM for generating and submitting requests) and data objects (DM for processing requests). The rollback mechanism based on time warp takes place at both TM and DM objects.

Virtual time incorporates several different data structures and control algorithms to produce optimistic processing. We use the Time Warp based implementation of virtual time in which rollback is the fundamental synchronizer in the system. The data structures include queues for storing messages, a unique process *id*, and a virtual time.

Before describing the VT synchronization mechanism in DDBS, let us discuss the structure of messages and objects that will be used within this scheme.

- **Messages**: A transaction request or response is represented by a message which contains the following information: the object's sender id, the object's receiver id, the virtual time (also called timestamp), the type of message, the content of the message (i.e., the text), and the sign. In our implementation, we use the current value of the system clock to generate timestamps, ensuring that no two timestamp values are generated during the same tick of the clock. In order to guarantee that each transaction has a unique timestamp, each virtual time has two components – the timestamp value and the unique process identifier, *Pid*. The type and text components have the query information. The possible message types are *read, write, read-response*, and *commit* messages. The sign field holds a bit indicating whether this is an antimessage or not, as required by the rollback mechanism [19].

- **Database Manager (DM) Internals**: To support the distributed processing and the rollback mechanism, the structure of the DM consists of the following: *Pid*, Local Clock, Permanent Value, Input Queue and Output Queue.
  The *Pid* is the unique process identifier for a node; the local clock is the current local virtual time of an item; the permanent value is the last committed write value; the input queue stores messages from the virtual past along with unprocessed messages in the manager's virtual future for an item; and the output queue keeps the antimessage copies of all *read-response* messages generated and sent by the manager.
  The DM continuously processes those messages that have already arrived in its input queue and stores copies of any messages it sends in an output queue. The local clock is set equal to the timestamp of the message being processed. The output queue and the virtual past portion of the input queue are required to support the rollback mechanism. Whenever a straggler (i.e., a message with timestamp less than the local virtual time of the receiving process) does arrive, a rollback occurs which means the Time Warp protocol restores the state of the object, cancels the side effects, and starts processing the affected objects forward again.
- **Transaction Manager (TM) Internals:** The TM consists of three primary components which hold the information required for the rollback mechanism and the distributed processing. These are: *Pid*, Local Clock, the Output Queue, and Input Queue (which will contain the messages sent from the users).
  The transaction manager adjusts its local virtual time by setting its local clock equal to the timestamp assigned to the current transaction. The output queue is ordered by timestamps and contains antimessage copies of all *write* messages generated and sent by the manager. Since the TM generates the *read* and *write* requests, no input queue is needed.
- **Virtual Clocks:** In the virtual time synchronized DDBS, each transaction manager has its own local virtual clock, and each item managed by the database manager has also its own virtual clock. The local virtual clock represents the virtual time of the message (or transaction) currently being processed.

4.1.1. Local Control Mechanism. Both database and transaction objects use the time warp paradigm [19]. Thus, they both need to perform common system tasks such as managing messages, clocks, queues, events, and rollbacks as discussed below.

- **Message Management**: In a Time Warp mechanism, since message passing is used so frequently, it is crucial that messages are managed efficiently. In our implementation, we choose the following paradigm: when a message for an item arrives at either database or transaction manager, it is placed in an input queue with two distinct parts. The first part contains unprocessed messages for the virtual future time, while the second part contains processed messages from the virtual past. Before a newly arrived message is placed in the input queue, the sign component is examined to determine whether the message is an event message (+ sign) or an antimessage (– sign) in order to handle them differently.
Upon arrival of an event message, the manager (i.e., DM or TM) places it in the input queue in the timestamp order. If the message's timestamp is in the manager's virtual past, the manager must rollback to a virtual time earlier than the timestamp. Execution begins at this earlier time, ensuring that all messages are processed in timestamp order. The rollback mechanism is invoked to undo all the work done after the timestamp. If the timestamp is in the manager's virtual future, no further action is taken since the message has been queued. Thereafter, the manager resumes processing messages in the input queue.
Upon arrival of an antimessage, the manager also places it in the input queue in timestamp order. This implies that some transaction managers have rolled back[3] and are undoing work that should not have been done. If the antimessage's timestamp is in the manager's virtual future, the corresponding message found in the virtual future part of the input queue should be removed. If the antimessage's timestamp is in the manager's virtual past, the manager must be rolled back to a virtual time earlier than the antimessage's timestamp. All the work done after the antimessage's timestamp must be undone. In either case, the antimessage will eventually annihilate its corresponding message. Since the only purpose of the antimessage is to seek out its corresponding event message, it is canceled along with any side effects it may have caused. The timestamps of the antimessage and its corresponding message are the same.
The output queue is used strictly to support the rollback mechanism. If a manager rolls back, antimessages may be sent out to cancel the original messages. The output queue contains timestamps ordered antimessages corresponding to the messages the manager has sent. All the work completed by a manager is conditional and could be rolled back.
As long as unprocessed messages are in the input queue, the database manager processes them. When a *read* message is processed, the input queue is searched for the latest, previous *write* messages. If no *write* is found, the value recorded in the database is used. A response message

[3] Note that the rollback is generated at the object level, i.e., transaction and data objects, and a transaction consists of a set of sub-transactions, i.e objects.

with this value is generated and sent to the sender object (or process). An antimessage copy is stored in the output queue in case a rollback is required later. In our implementation, a *write* message is simply marked as 'processed'. The physical *write* operation must be delayed until it can be committed. This action will be discussed later. Recall that the *read* messages are not rolled back since there is no causal relationship between the *read* operations.

- **Clock Management**: The synchronization of clocks in a virtual time environment is not typical of other distributed database methods. The local virtual clock for a manager (TM or DM) is allowed to move ahead or behind in virtual time. The clock moves ahead by processing messages in its input queue while it moves behind due to rollback by an incoming message timestamped in the virtual past. The local clock for the transaction manager always moves forward as transactions are completed. The local clocks of all managers are at best, loosely synchronized. Overall, it is expected that the local virtual clock advances ahead in time while occasionally falling behind.

4.1.2. DM and TM Rollback Mechanisms. A key aspect of the time warp based mechanism is that it uses rollback as the principal synchronization primitive. In this approach, objects proceed with no delay and rollback whenever out-of-order messages arrive. The appeal of such an optimistic method depends on the efficiency of the rollback mechanism employed. Several strategies have been suggested in the literature [12, 17, 19]. In our implementation, we use the following rollback schemes for the database and the transaction managers.

The rollback scheme for the DM handles late arrival *read* messages, *write* messages, and *write* antimessages, each of which is handled differently. Our approach follows a method based on optimized semantics [2] as opposed to traditional rollback mechanisms. The basic idea is not to rollback a manager just because a late message arrives. Instead, the message is processed at the current virtual time if the operation involved does not violate the computational correctness. A typical example is a late arriving *read* message.

Let us now describe the rollback scheme for both DM and TM managers.

Upon receipt of a *read* message, the DM inserts it into the input queue and treats it as a normal *read* message. If a *write* message is received, it is inserted into the input queue and marked as 'processed'. On the other hand, if a *write* antimessage is received, then its corresponding *write* from the input queue is removed. In both cases, a *read-response* antimessage is sent for any previously affected *read-response* messages.

For the late arrival *read* messages, the receiver process (object) will find the latest, previous *write* and return the value recorded. When a *write* message is received, any processed *read* messages with timestamps larger than the late *write* must be reprocessed. This is done by canceling the *read-response* messages and sending out a new response message. If a *write* antimessage is received, the corresponding *write* messages must be annihilated. Any *read-response* message which was satisfied with this value must be canceled, too. The net effect after rollback must be to leave the database in a state as if the late message had arrived in order. Antimessages may be sent to accomplish this, but are not always necessary.

Since a transaction may consist of a set of sub-transactions, individual actions within a transaction are only rolled back when conflict occurs. The rollback mechanism for the TM handles only *read-response* antimessages because these are received exclusively. If there is no *write* message sent then only the new *read-response* values are recorded. Otherwise the *read-response* values are recorded, and the write messages need to be canceled. Thus, *write* antimessages will be transmitted for all writes sent, the *write* messages will be sent. It is assumed the *reads* have no interdependencies.

4.1.3. Global Control Mechanism. The success of the virtual time protocol is embedded in global control issues. These include measuring global progress, managing memory, controlling message flow, committing irreversible actions, and taking snapshots. The focal point of all these issues is the *global virtual time* (GVT) computation [19]. This is a very important concept in virtual time because of its global impact, especially on the global progress.

- **Global Clock**: Every manager[4], in a virtual time system has its own local clock. Individual clocks contain different virtual times and may progress forward or backward in time. No single local clock can be used as an indicator for the global progress of the system. Therefore, a global clock containing the GVT is needed. The GVT always progresses forward at an unpredictable rate with respect to real time. This is the first requirement to assure the handling of the problems present in any Time-Warp based mechanism [16, 19]. The GVT also serves as an absolute lower bound beyond which no process can roll back, and is typically used to perform fossil collection of obsolete messages and states. It is necessary to compute a GVT estimate periodically in order to do garbage collection. We make use of a simple algorithm where a predefined node collects the local virtual times and periodically computes the GVT.
- **Memory Management**: Experience with Time Warp over the past years has revealed two fundamental problems with this paradigm, namely memory management resulting from the necessity to save states and unstable behavior. This memory management problem can be tackled with the help of cancelback protocols and the use of artificial rollback [16, 12, 17].

  In our distributed database model, each manager stores the history information in case of rollback. However, a manager does not need to store history from the beginning of the computation if it cannot possibly roll back that far into the past. For efficient memory management,

[4] In the sequel, we will use manager to indicate both TM and DM managers, unless specified otherwise.

we choose a simple fossil collection algorithm. Any message in the input queue with a timestamp less than the GVT is discarded. If the message is a *write*, then its value is recorded as a permanent value. Similarly, any message in the output queue with a virtual send time less than the GVT is discarded. The discarded memory is recovered and reallocated to future requests.

- **Commitment**: The nature of rollback in Time Warp restricts a class of operations from being performed immediately. These operations include irreversible actions. In our case, the *write* operation to permanent storage falls into this class. A Time Warp system requires that no irreversible action be committed until it can be proven correct. To enforce it, a *write* operation must be buffered (not committed) until the virtual time associated with it is less than the GVT. The delay due to buffering is not considered a major drawback of virtual time, but specific applications could suffer from it. One of our objectives in this paper is to determine how long this delay is and whether it could adversely affect the user response time.

4.2. Costs Involved. Memory requirements, message traffic, and execution time contribute to the overall cost of a virtual time system. Each of these can be manipulated by the virtual time designer to tune the system for optimal performance. The most critical parameter is the frequency of GVT computation, since it affects the response time, throughput, and memory requirements. A high frequency leads to a better response time and memory management, but requires more processing time and increases message traffic. On the other hand, a low frequency has the opposite effect. In this paper, the GVT for every test run was individually tuned to achieve maximum performance.

5. Synchronization by MultiVersion Timestamp Ordering. MultiVersion Timestamp Ordering (MV) is a well known concurrency control mechanism to achieve DDBS synchronization. We implement an MV protocol to serve as a basis for comparing our results obtained with the VT protocol. For the sake of completeness, we outline below the underlying concepts of an MV protocol.

The basic idea is to keep several versions (values) of an item. When a transaction requires access to an item, an appropriate version is chosen to maintain the serializability of the concurrently executing schedule, if possible. The MV scheme processes requests on a first-come first-serve basis. Operations on data items are translated into operations on versions to make it appear as if it had processed the operations in a timestamp order on a single version of the database. Several variants of MV protocol have been reported in the literature. In this paper, we use the multiversion algorithm due to Bernstein and Goodman [3].

In our implementation, the MV protocol requires a database manager and a transaction manager. The DM consists of several buffers and lists to accomplish synchronization. It relies on abortion as its fundamental synchronizer. The TM generates transactions and sends requests to the database managers. These requests consist of *read, Pre-write*, and *write* operations. Each transaction is assigned a unique timestamp. If a transaction aborts, it is resubmitted with a new timestamp. This cycle repeats until the transaction is successfully completed. Timestamps in this system are also based on real-time, but we do not consider clock synchronization.

The TM requires *read, Pre-write*, and *write* phases to submit a transaction. In the *read* phase, *read* requests are sent which are blocked until a response is received. The *Pre-write* and *write* phases are necessary to support a two-phase commit protocol.

In an MV system, there is no clock synchronization. Hence, it is possible for a message to arrive out of timestamp order. When a straggler message arrives, the transaction manager may need to abort the transaction to resolve the time conflicts. If a transaction is aborted, the TM will assign a new (and larger) timestamp to the aborted transaction.

The DM rejects a *Pre-write* message explicitly by sending an *abort* message back to the TM. Upon receipt of an *abort* message, the TM must cancel all the requests involved in this transaction and restart it. This process could be repeated if many conflicts are detected. This is a primary weakness of an MV synchronization protocol.

A *write* message is *committed* as soon as it is received by the database manager. A commit guarantees the permanence of the effects of a transaction. The transaction manager knows that this message is expected since it is preceded by a *Pre-write* message. The time between these two messages is considered as a commitment delay, which is not the same as the delay in a virtual time synchronizer.

System tuning is also important in the MV system. The most crucial parameter is the timeout period which was obtained empirically in our experiments. This value determines how long the transaction manager must wait for a possible abort response after issuing all of its *Pre-writes*.

6. Performance Study. We conducted our experiments on a bus-based message passing environment consisting of 12 Sun SPARC workstations. The machines were connected into a local area network (LAN) via ethernet. The ISIS V2.1 toolkit [7] for distributed and fault-tolerant programming was used to implement the MV and VT synchronization protocols. The ISIS is built on top of the UNIX operating system, and developed mainly for easy coding of distributed and parallel applications. In our models, ISIS CBCAST protocol was primarily used as a message passing facility. The language used is C.

Virtual Time Database Testbed . We used a rollback mechanism for both the TM and DM. A transaction manager issues a set of *read* requests followed by a subset of *write* requests. Since it issues only one transaction at a time, it could not be rolled back beyond that transaction. We assumed that a *read-response* message that is rolled back does the *writes*, and it does not affect the other members of the *read* set. Furthermore, once the rollback started at either manager, it was allowed to finish without any interruption. Normally, the rollback operation would be allowed to be interrupted in case a message further back in time needed to processed. This diverged from normal rollback, but was easier to implement and orient towards ISIS message delivery. The rollback mechanisms took this into consideration.

The history information stored for rollback included all entries in the output queue and a portion of the input queue. One structure not needed in this implementation of virtual time was the state queue which contains the checkpoint information.

MV Testbed. As mentioned earlier, periodic maintenance has to be performed on the *read* and *write* request lists. In our MV testbed, this was accomplished by periodically removing old entries which no longer affect the synchronization.

In our experiments, some assumptions were made about the nature of the transactions and their submissions since they have an impact on the performance of the DDBS. These models were similar to the ones suggested in [1, 11, 20]. No blind *writes* were allowed. A transaction could not issue a *write* operation to a database item without first reading it. The transaction manager generated the user requests randomly and issued them to the database managers. Only *read/write* synchronizations were considered. The control variables for this model included the transaction/database sizes, the inter-transaction delay, and the update probability.

A simple database model was used. Each database manager was responsible for a specified number of items. Whenever a transaction requested a *read* or *write* access to an item, it was delayed 50 milliseconds to simulate the physical I/O operations.

6.1. Experimental Results. In our model, we made use of several database and transaction sizes, and we varied the number of processors from 4 to 12. We present our results in terms of response time, the message traffic, and the throughput. We present our results below in the form of graphs of the response time (in seconds), message traffic and throughput of the model as a function of database and transaction size as well as the number of processors employed in the model.

The *response time* is defined as the average amount of clock time per processor required for a transaction manager to successfully submit a fixed number of transactions. The *message traffic* is calculated by counting every message sent by both DM and TM managers, and as in [20], the *throughput* in committed sub-transactions is defined as $(P * N_t)/(\text{total response time of all processors})$, where P is the number of processors used, and N_t is the number of transactions. In our experiments, $N_t = 500$.

The following experimental data were obtained by averaging several trial runs. Every test run required between 8 and 50 minutes of actual time to complete all experiments. The average length of the experiments was 24 minutes. Note that the results presented here are tempered with the overhead required to execute ISIS.

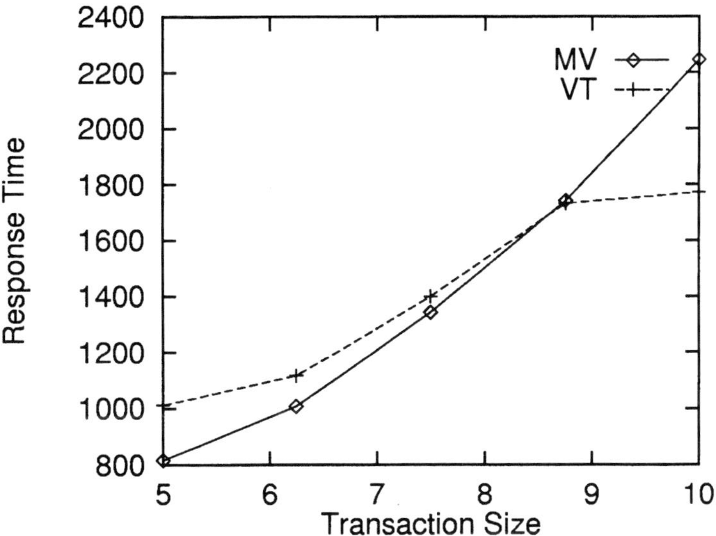

FIG. 6.1. **Response Time vs. Transaction Size**

6.1.1. Effect of Transaction Size. We first analyze the effect of the transaction size on the performance of the DDBS synchronized by virtual time. To this end, several simulation tests were run on 8 processors. We varied the transaction size from 5% to 10% of the database size. All *read* requests were updated with probability 25%. Figure 6.1 portrays the values obtained for the response time of the simulation of DDBS for both VT and MV protocols. As seen from the graphs, the VT scheme exhibits a better response time when compared to the MV scheme when the transaction size is larger than 8.5%. The MV performs better only for a small transaction size. This is due to the intrinsic properties of virtual time. Choosing to undo work as in the VT scheme rather than aborting it as in the MV scheme has appeal since abortion is a more drastic and costly operation.

Figure 6.2 shows the message traffic as a function of the transaction size. The results demonstrate that good response time is strongly affected by high overhead in the synchronization strategies (i.e., VT and MV). The graphs show that the message traffic increases with the transaction size for both VT and MV schemes. We also observe that the VT scheme induces less traffic than MV when the transaction size is larger. This is due to the fact that there are less rollback operations with VT, and MV has more abortion operations since rescheduling the transaction request later induces another message.

Figure 6.3 depicts the throughput for both VT and MV protocols. In both schemes, the throughput decreases with the increase in the transaction size. It is also important to note that the VT scheme exhibits much larger throughput than the MV scheme when the transaction size exceeds 9. We believe that this

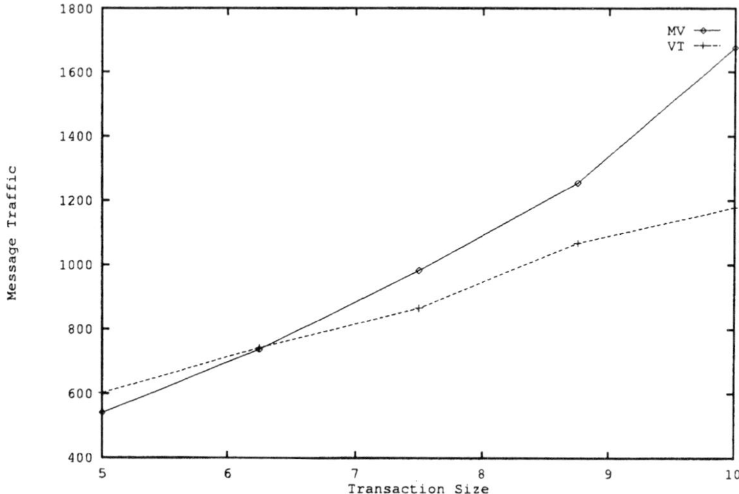

FIG. 6.2. **Traffic Message vs. Transaction Size**

is due to the fact that the recoverability (as in VT) is an easier condition to enforce than the serializability which is identified by the chronological order of timestamps of the concurrent transactions (as in MV) and the maintenance of the data item versions.

6.1.2. Effect of Database Size. Let us analyze the effect of the database size on the performance (response time and throughput) of the DDBS synchronized by virtual time. Once again, we used 8 processors and varied the database size from 1 to 20. The transaction size was set to 7% of the database size. All *read* requests were updated with a probability 25%. The inter-transaction delay[5] was set to zero. Figure 6.4 presents the response time as a function of database sizes for both synchronizers VT and MV. The results show that MV and VT exhibit about the same response time for database size less than 12. When the size is increased from 12 to 20, VT responded faster than MV. We observe about 25% reduction in the response time for a database size of 20.

Figure 6.5 shows the message traffic as a function of database sizes. Once again, the message traffic is higher with MV when compared to VT. For instance, we observe about 30% more traffic with MV than VT. This is due to the properties of VT and the fact that recoverability is much easier to satisfy than to enforce serializability.

In Figure 6.6, we present the values of the throughput for both strategies (VT and MV). The results again mirror the curves obtained for the response

[5] Later, we study the effect of inter-transaction delay on the response time and the throughput.

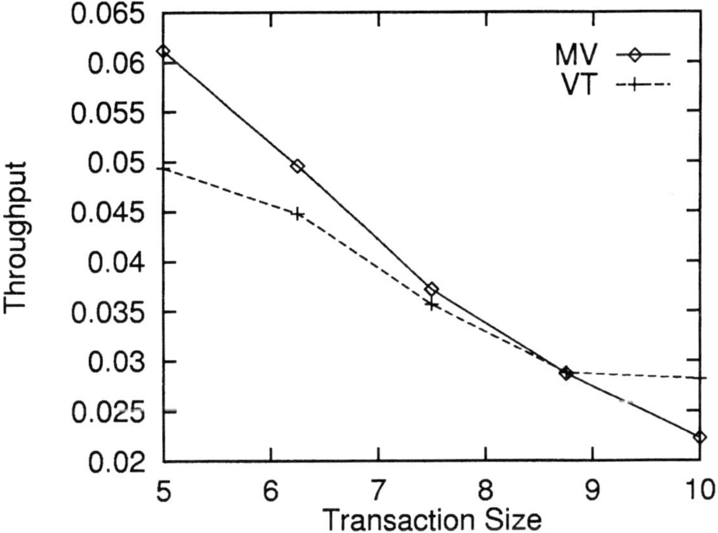

FIG. 6.3. **Throughput vs. Transaction size**

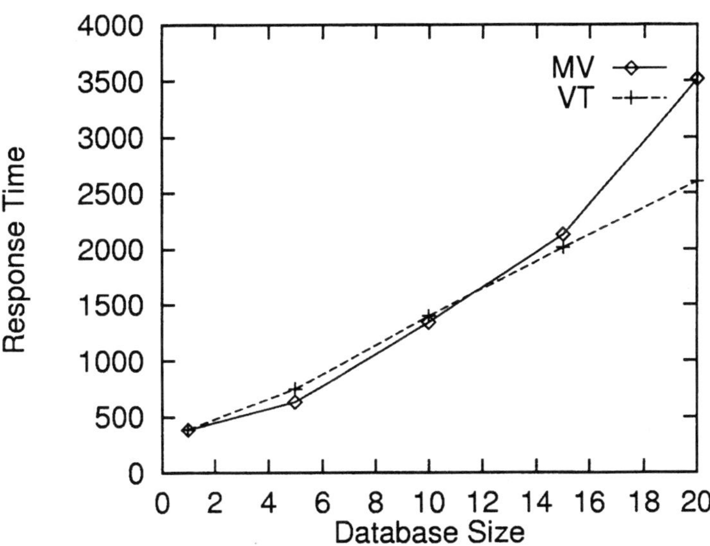

FIG. 6.4. **Response Time vs. Database Size**

time. The throughput decreases significantly as we increase the database size.

6.1.3. Effect of Inter-Transaction Delay. In the following, we report on the experiments which were carried out to study the effect of the inter-transaction delay on the overall performance of our DDBS model. As illustrated in Figure 6.7, the VT method exhibits a better response time when compared to

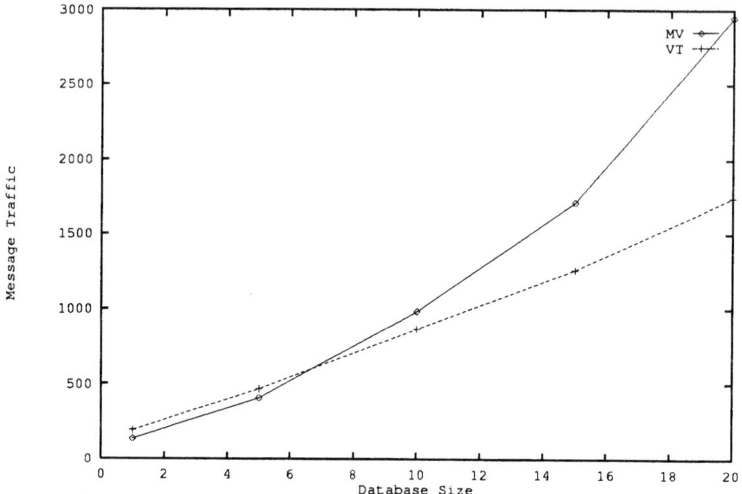

FIG. 6.5. **Message Traffic vs. Database Size**

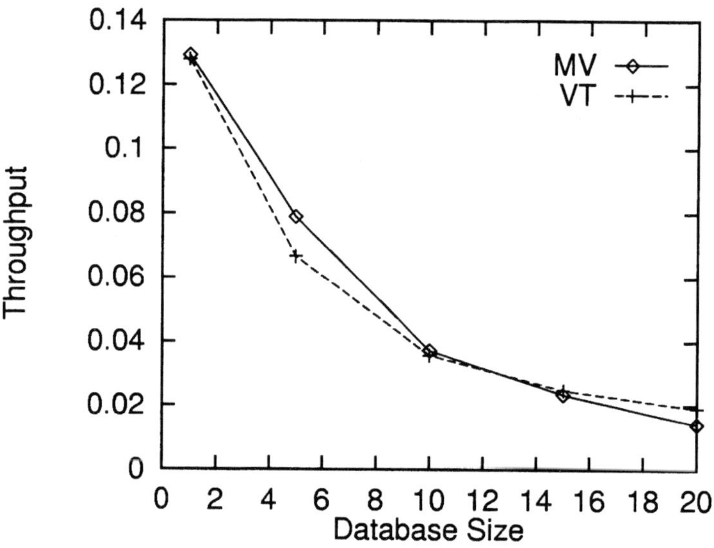

FIG. 6.6. **Throughput vs. Database Size**

the MV scheme as we increase the inter-transaction delay. Figure 6.8 portrays the results obtained for the message traffic when we vary the inter-transaction delay. As we can see, the message traffic is higher with MV than with VT. Figure 6.9 demonstrates that VT exhibits a better throughput when compared to MV and when we vary the inter-transaction delay using the same DDBS model. This is basically due to the fact that the recoverability is less costly than maintaining the serializibility.

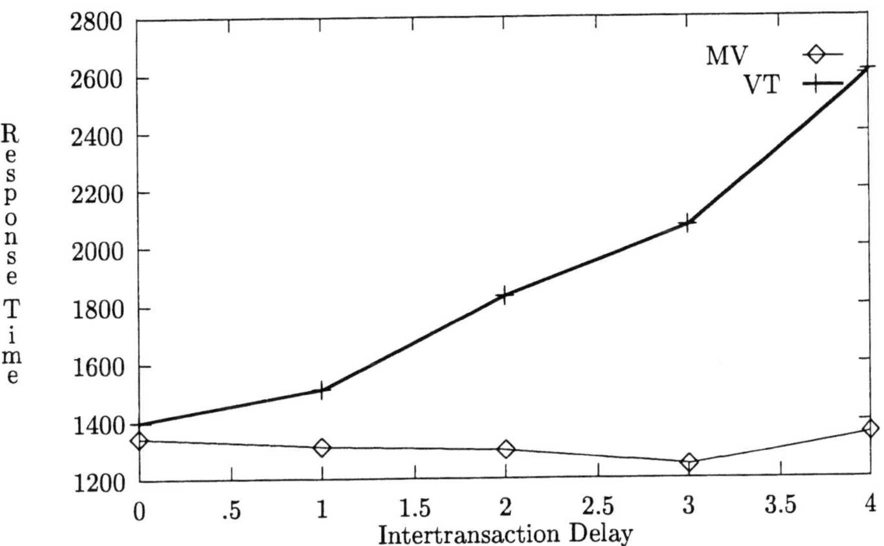

FIG. 6.7. **Response Time vs. Inter-transaction Delay**

6.1.4. Effect of Processors. Finally, in order to reduce the response time of DDBS, we ran experiments varying the number of processors from 4 to 12. The results obtained for the response time and throughput are depicted in Figures 6.10 and 6.11. As expected, as we increase the number of processors, the response time increases for both the synchronization schemes. Furthermore, the VT scheme performs better than the MV scheme. For example, 25-30% reduction in the response time is observed with the VT scheme as compared with the MV scheme. Figure 6.11. shows that the throughput decreases as we increase the number of nodes for both schemes. Here again, the curves obtained for the throughput mirror the results for the response time.

7. Conclusions. The behavior of database concurrency control schemes has been an area of extensive research. However, not much has been reported on the performance of distributed databases on a network of workstations. In this paper, we have described an implementation of the virtual time method as an alternate concurrency control scheme for DDBS using a LAN connected collection of Sun SPARC machines.

Our experimental results indicate that careful implementation of virtual time using the time warp synchronization mechanism is a viable technique,

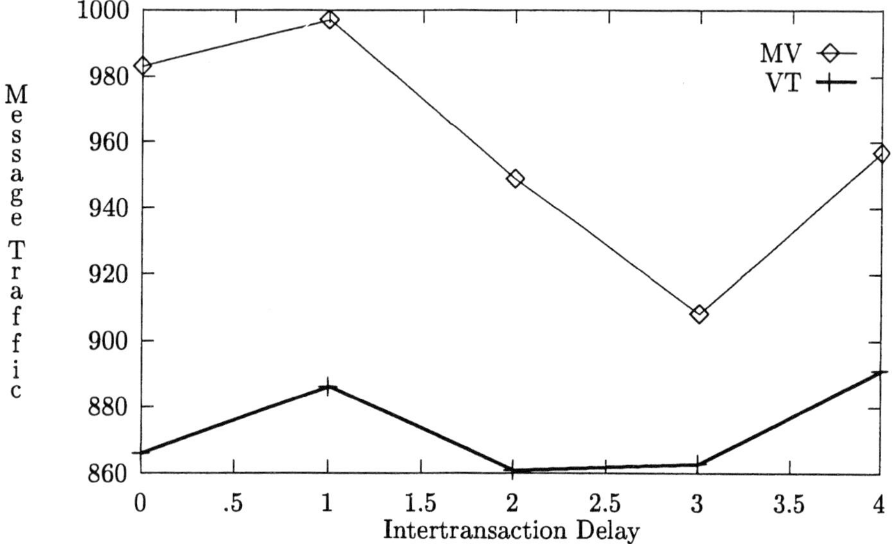

FIG. 6.8. **Message Traffic vs. Inter-transaction Delay**

and improves the performance (response time and throughput) of distributed databases. We studied the effect of transaction size, database size, and inter-transaction delay on the overall performance of the DDBS. The results demonstrate that the VT protocol outperforms the MV scheme as the transaction size increases. We also observe about 25-30% reduction in the response time using the VT method over the MV scheme. This is due to the fact that a recoverability operation is a much easier condition to enforce than serializability and an undoing operation is less costly than an abortion. The results also show that adding inter-transaction delay to a VT scheme will only help to degrade the performance of the system when compared to an MV one. As regards to scalability, as we increase the number of processors, VT outperforms MV.

The results presented in this paper are very encouraging. We believe that more research needs to be done in the area of distributed databases on clusters of workstations since cluster/network of workstations are gaining more popularity and easily available.

Acknowledgments. The authors would like to thank the anonymous authors whose valuable comments largely improve this paper.

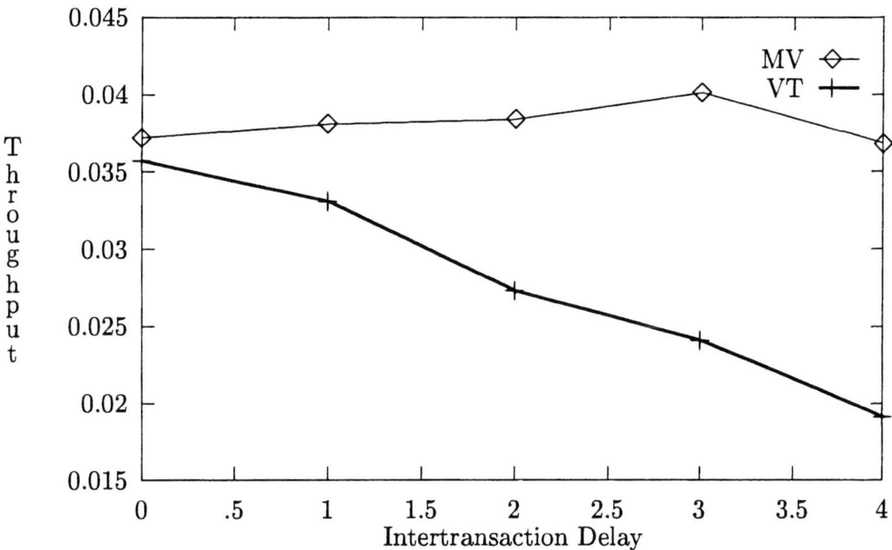

FIG. 6.9. **Throughput vs. Intertransaction Delay**

REFERENCES

[1] R. AGRAWAL AND M.J. CAREY, *Concurrency Control Performance Modeling: Alternatives and Implications*, *ACM Transactions on Database Systems*, Vol. 12, No. 4, 1987, pp. 609–654.

[2] BADRAINATH, AND K. RAMAMRITHAM, *Semantic-Based Concurrency Control: Beyond Computativity*, TODS, 17:1, 1992.

[3] P.A. BERNSTEIN AND N. GOODMAN, *Concurrency Control in Distributed Database Systems*, *ACM Computing Surveys*, Vol. 13, No. 2, 1981, pp. 185–221.

[4] BOUKERCHE A., *Time Management in Parallel Simulation*, Chapter in High Performance Cluster Computing, Prentice Hall, Vol. 2, Ed. B. Rajkumar, 1999, pp. 375–394.

[5] A. BOUKERCHE, AND T. W. TUCK, T3C: A TEMPORALLY CORRECT CONCURRENCY CONTROL ALGORITHM FO DISTRIBUTED DATABASES, Proc. of the 8th IEEE MASCOTS, 2000, pp. 155–163.

[6] A. BOUKERCHE, S. K. DAS, A. DATTA, AND T. LEMASTER, *Implementation of a Virtual Time Synchronizer for Distributed Databases on a Cluster of Workstations*, Proc. of IEEE International Parallel and Distributed Processing Symposium, 1999, pp. 733–737.

[7] K. BIRMAN, R. COOPER, T. JOSEPH, K. MARZULLO, M. MAKPANGOU, K. KANE, F. SCHMUCK, AND M. WOOD, *The ISIS System Manual, Version 2.1*, The ISIS Project, September 1990.

[8] M.J. CAREY AND M. LIVNY, *Distributed Concurrency Control Performance: A Study of Algorithms, Distribution, and Replication*, *Proceedings of the Fourteenth International Conference on Very Large Data Bases*, 1988, pp. 13–25.

[9] K. M. CHANDY, AND J. MISRA, *Asynchronous Distributed Simulation via Sequence of Parallel Computations*, *CACM* 24(3), Apr. 1981, pp. 198–206.

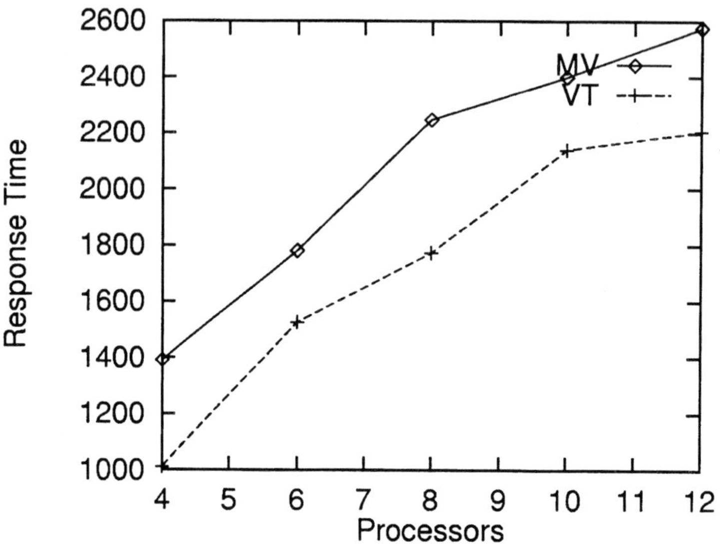

FIG. 6.10. **Response Time vs. Number of Processors**

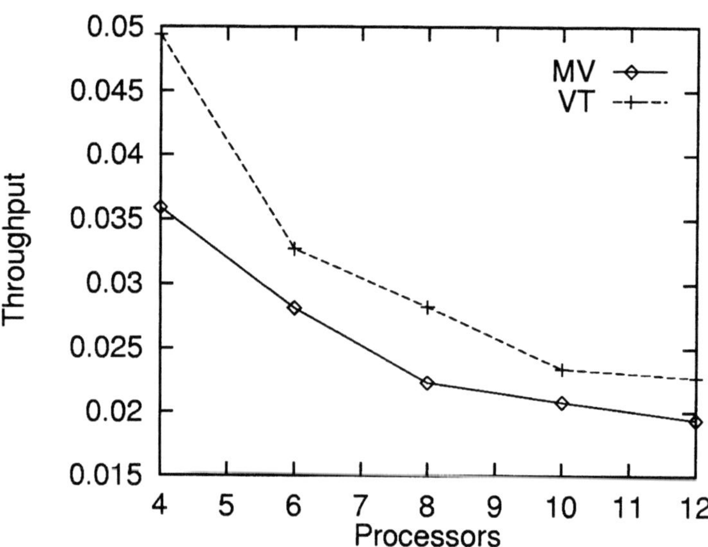

FIG. 6.11. **Throughput vs. Number of Processors**

[10] M. CHANDY AND L. LAMPORT, *Distributed Snapshots: Determining Global States of Distributed Systems* ACM Transactions on Computer Systems, Vol. 3, No. 1, 1985, pp. 63–75.

[11] U. DAYAL, AND R. LADIN, *A Transaction Model for Long Running Activities*, VLDB, 1991.

[12] S. DAS, AND R. FUJIMOTO, *A Performance Study of the Cancelback Protocol for Time*

Warp, ACM/IEEE PADS'93, pp. 135–142.

[13] B. C. DESAI, *Database Systems*, West. Pub., 1990

[14] R. ELMASRI, S. NAVATHE, *Fundamentals of Database Systems*, Addison Wesley, 1994.

[15] R. M. FUJIMOTO, *The Virtual Time Machine*, Proceedings of the ACM Symposium on Parallel Algorithms and Architectures, June 1989, pp. 199–208.

[16] R. M. FUJIMOTO, *Parallel Discrete Event Simulation*, CACM, Vol. 33, No. 10, Oct. 1990, pp. 30–53.

[17] A. GAFNI, *Rollback Mechanisms for Optimistic Distributed Simulation Systems*, Proceedings of the SCS Multiconference on Distributed Simulation, Vol. 19, No. 3, July 1988, pp. 61–67.

[18] D. JEFFERSON AND A. MOTRO, *The Time Warp Mechanism for Database Concurrency Control*, International Conference on Data Engineering, 1986, pp. 474–481.

[19] D. R. JEFFERSON, *Virtual Time*, ACM Transactions on Programming Languages and Systems, Vol. 7, No. 3, 1985, pp. 404–425.

[20] C, HASS AND WEIKUM, *A Performance Evaluation of Multi-Level Transaction Management*, in VLDB, 1991.

[21] W. H. KOHLER, *A Survey of Techniques for Synchronization and Recovery in Decentralized Computer Systems*, ACM Computing Surveys, Vol. 13, No. 2, June 1981, pp. 149-183.

[22] C. U. ORJI, L. LILIEN, AND H. HYSIAK, *A Performance Analysis of an Optimistic and a Basic Timestamp-Ordering Concurrency Control Algorithms for Centralized Database Systems*, IEEE International Conference on Data Engineering, 1998, pp. 64–71.

[23] C. H. PAPADIMITRIOU AND P.C. KANELLAKIS, *On Concurrency Control by Multiple Versions*, ACM Transactions on Database Systems, Vo. 9, No. 1, March 1984, pp. 89–99.

[24] A. WITKOWSKI, *Performance Evaluation of Timestamp Driven Databases*, PhD thesis, Department of Computer Science, University of Southern California, September 1985.

APPLYING MULTILEVEL PARTITIONING TO PARALLEL LOGIC SIMULATION*

SWAMINATHAN SUBRAMANIAN[†], DHANANJAI M. RAO[‡], AND PHILIP A. WILSEY[§]

Abstract. The size and complexity of hardware systems motivates the use of simulation for their study and analysis. Parallelization techniques are often employed to meet the memory and computational requirements for simulating large hardware designs. Furthermore, partitioning the design for parallel simulation is vital for achieving acceptable simulation throughput. This paper presents the design and implementation of a new partitioning algorithm based on a multilevel partitioning heuristic. The multilevel algorithm attempts to balance the load, maximize concurrency, and reduce inter-processor communication in three phases to improve performance of the parallel simulator. The paper presents the algorithm and reports results from of our empirical studies to compare performance of the multilevel algorithm with other partitioning techniques. A generic partitioning infrastructure was developed and integrated into the WARPED parallel simulation framework to ease design, implementation, and study of different partitioning algorithms. The design issues involved in the development of the generic partitioning infrastructure are discussed and the techniques employed to integrate several partitioning algorithms into this framework are also presented in the paper. The experimental results obtained from our benchmarks indicate that the multilevel algorithm yields better partitions than other partitioning algorithms included in the study.

Key words. Partitioning, Parallel Simulation, Logic Simulation, Circuit Simulation

AMS subject classifications. 68U20, 68W35

1. Introduction. The design and analysis of any system is governed by the parameters that influence the system. Often, analysis involves the study of interactions between such parameters. A complete analysis of a system that is influenced by n parameters involves the study of 2^n interactions. In the case of smaller systems, analytical methods or prototyping techniques can be applied. Such techniques provide fast and accurate results if the number of parameters is small. On the other hand, a combinatorial increase in the number of interactions is experienced in the design and analysis of large and complex systems, rendering analytical and prototyping techniques costly, tedious, and time consuming. For such systems simulation techniques can be employed to provide fast, feasible, and inexpensive solutions. Accordingly, discrete event simulators are often used to analyze and evaluate hardware circuit designs. Unfortunately, the performance of sequential simulators deteriorates with increases in size and complexity of the circuits involved. Parallel discrete event simulators can be used to alleviate this problem by distributing the workload for simulation among several proces-

*Support for this work was provided in part by the Defense Advanced Research Projects Agency under contract DABT63–96–C–0055.

[†]Experimental Computing Laboratory, Cincinnati, OH 45221–0030 (swamin@cadence.com).

[‡]Experimental Computing Laboratory, Cincinnati, OH 45221–0030 (dmadhava@ececs.uc.edu).

[§]Experimental Computing Laboratory, Cincinnati, OH 45221–0030 (philip.wilsey@ieee.org).

sors [3]. However, parallel simulators can suffer from numerous overheads such as inter-processor communication and workload imbalance. Such overheads become significant when simulating large circuits. Hence, dividing and assigning the workload equally across the processors such that the overheads are minimized is critical for maximum performance. Hence, *partitioning*, the process which computes such an assignment, is a key factor in achieving speedup in parallel simulations [10, 31, 35].

Traditionally, partitioning techniques were designed to exploit either the parallelism inherent in (i) the simulation algorithm, or (ii) the model being simulated [27]. The amount of parallelism that can be gained from the former techniques is limited by the algorithm used for simulation. The latter techniques attempts to improve performance by dividing the model being simulated across processors. Hence, during simulation, the workload is distributed and the concurrency and parallelism in the model are exploited. The success of the latter techniques is bounded by the amount of parallelism inherent in the model and the number of processors available for simulation. The partitioning studies presented in this paper fall under the latter category. However, the parameters for the partitions algorithms must also encompass the characteristics of the underlying simulation kernel in order to yield efficient partitions. In this paper, the partitioning studies, are conducted within the context of parallel simulations that employ the Time Warp [16] synchronization strategy. In a Time Warp simulation, the parallel processes are asynchronous and their performance is strongly influenced by Time Warp overheads such as rollbacks and state saving/restoration overheads [13, 16]. To fully benefit from the optimistic behavior of Time Warp, these overheads must be minimized. In general, these overheads depend upon the amount of communication, the degree of concurrency, and the load balance in the simulator. Consequently, an effective partitioning scheme for Time Warp simulators must simultaneously optimize all three factors to achieve maximum performance [1]. Hence, the multilevel partitioning algorithm, developed as a part of the partitioning studies, concentrates on improving the performance by maximizing concurrency, minimizing inter-processor communication, and balancing the processor workload based on the circuit being simulated. The new multilevel approach to partitioning, developed in this paper, attempts to optimize the aforementioned factors by The complexity of the multilevel algorithm is linear with respect to the interconnectivity in the circuit making it a fast linear time heuristic.

The partitioning studies reported herein were conducted to improve the performance of the SAVANT parallel VHDL simulation framework [36]. To ease design, implementation, and study of different partitioning algorithms, a generic partitioning framework was developed and integrated with SAVANT. The framework provides an Application Program Interface (API) for developing partitioning algorithms and provides necessary insulation between the partitioning techniques and the simulation core. Several partitioning algorithms, including the new multilevel approach, were implemented using this framework to analyze and study their advantages and shortcomings. This paper presents the issues involved in the design and implementation of different partitioning algorithms along with the experiments conducted to evaluate their effectiveness. In §2, a brief overview of partitioning is presented. A brief summary of several previously developed partitioning techniques is presented in §3. Section 4 describes the multilevel partitioning algorithm developed as a part of this study. A brief description of the experimental framework and the issues involved in the design and integration of the partitioning algorithms are presented in §5. The results of our experiments are presented in §7. Section 8 presents some concluding remarks along with pointers to future work.

2. Partitioning. Partitioning is driven by the notion of *divide and conquer*; instead of assigning a task to a single sequential processor, the task is divided among several processors that concurrently execute the sub-tasks to speed up execution of the task. Therefore, partitioning logic circuits across processors presents an interesting avenue to improve the performance of logic simulations. This section introduces concepts related to partitioning and develops a graph theory basis to view the problem of partitioning in the graph theoretic domain.

2.1. Classification of Partitioning. Partitioning schemes can be generally classified as either (i) static or (ii) dynamic. This classification roughly denotes when the partitioning mechanism is invoked. In a static partitioning scheme, partitioning is executed prior to simulation and the resulting assignment is fixed throughout the entire duration of the simulation. Static schemes sometimes use history information from previous simulation runs to help make partitioning decisions in subsequent partitioning activities. In contrast, dynamic partitioning schemes begin with an initial assignment and then routinely refine that partition throughout execution of the simulation. Dynamic schemes suffer from the migration overhead resulting from a change in assignment during simulation, while static schemes lack the capability to adjust to workload imbalance during simulation. Our studies focus on static partitioning and all the schemes for partitioning discussed in this paper are static.

2.2. Factors that influence Partitioning. A number of factors influence the performance of any static partitioning scheme. Since the assignment is fixed prior to simulation execution, an effort to identify such factors and predict their effects becomes important. It is well known that the three main competing factors that need to be addressed while partitioning for parallel logic simulation are [1, 10, 21, 35]:

(i) **Concurrency**: This factor is dependent on the activity of the logic gates during simulation. If an equal number of gates are active in all partitions at all simulation instances, maximum concurrency is achieved.

(ii) **Communication**: Message passing between partitions to either synchronize or send events gives rise to communication. Ideal partitioning for communication distributes the gates across processors such that no messages are sent across processors. In reality, such an assignment is not possible and so the goal becomes one of minimizing communication and related activities.

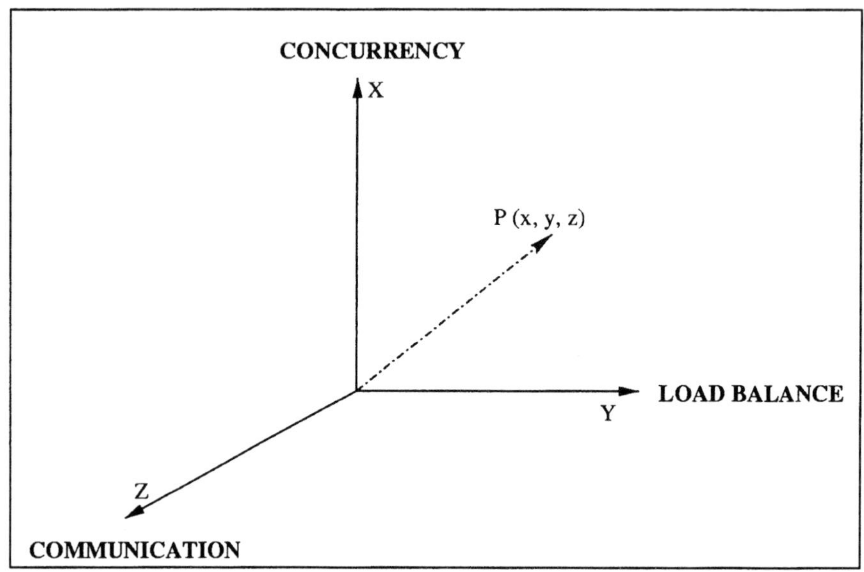

FIG. 2.1. *Factors that influence Partitioning*

(iii) **Load Balance**: Perfect load balancing requires that every partition have an equal load distribution at all times. This will maximize the performance of the simulator. However, a perfect load balance is practically impossible. Hence, the goal is to achieve as much load balance as possible.

Figure 2.1 illustrates the problem of partitioning in a 3-dimensional space; concurrency, load balance, and communication form the axes of this 3-D space. A point $P(x, y, z)$ in this space denotes a solution to the partitioning problem with concurrency, load balance, and communication values of x, y and z respectively. For example, if only a single partition is used for simulation, the communication is reduced to zero (there is no communication across partitions); but it results in zero concurrency, since the workload is not distributed and the execution proceeds in a sequential manner on a single processor. Similarly, if the entire workload is distributed with one logic gate per partition, the concurrency and the communication reach their maximum, since every task is allocated a processor and every event message must cross a partition boundary. Ultimately the goal of any partitioning algorithm is to *minimize communication, maximize concurrency, and optimize load balance* [1, 21, 35].

Graph theoretic representations have shown to be convenient for the study and analysis of partitioning algorithms and are widely used [1, 21, 35]. Most partitioning algorithms that are applied to digital logic circuits require a graph representation of the input circuit to be simulated. This graph is called the *circuit graph*. The circuit graph is represented as a directed graph $G = (V, E)$, where V forms the vertex/node set and E, the edge set of the graph. The vertices denote logic gates and edges represent *signals* [15] that interconnect these logic

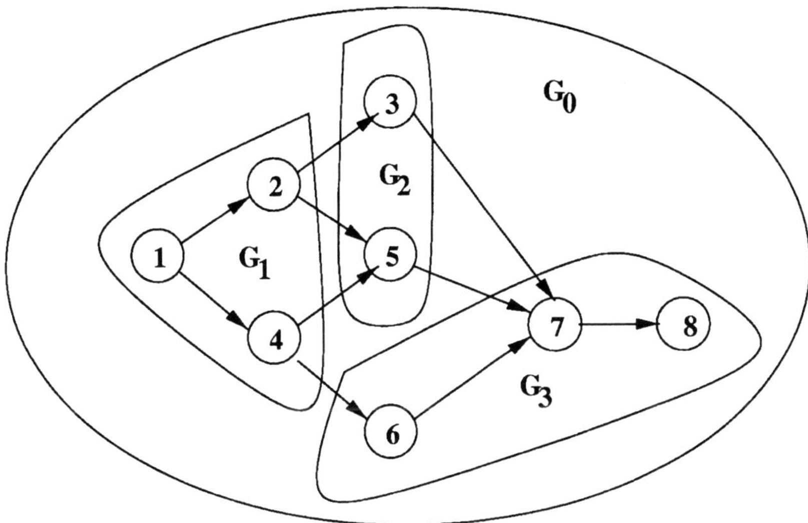

FIG. 2.2. *A Partitioned circuit graph*

gates (Figure 2.2). A partition of a circuit graph $G_0 = (V_0, E_0)$ is then a set of smaller sub-graphs G_1, G_2, ... G_k, where each $G_i = (V_i, E_i)$ $(1 \leq i \leq k$, k is the number of partitions) has its own vertex and edge set. The sub-graphs must satisfy the following relation:

$$\forall_{(V_i, V_j)}(V_i \subset V_0) \text{ and } V_i \cap V_j = \phi, \ (1 \leq i, j \leq k)$$

The edges that run across partitions belong to a special edge set called the *cutset*. The cutset represents the number of signals, and therefore, the number of messages that flow across partitions during simulation. In Figure 2.2, an example partitioning of an original graph G_0 into three sub-graphs G_1, G_2 and G_3 is illustrated.

3. Related Work. Many partitioning schemes based on graph theoretic approaches have been designed to the address various issues related to concurrency, communication and load balancing in different flavors. The basic idea behind the partitioning algorithms is to reduce communication and simultaneously balance load and maximize concurrency. The following subsections present a brief description of some of the partitioning algorithms that have been developed and investigated for parallel logic simulation.

3.1. Random Partitioning. Random partitioning is one of the simplest strategies to partition logic circuits. The gates in the circuit graph are randomly assigned to processors. Load balancing is maintained by distributing approximately equal number of gates across partitions. The obvious drawback of random partitioning is that it cannot guarantee reduction in communication or improvement in concurrency. Effective load balancing can be achieved using random assignment. However, communication can become a major bottleneck during simulation. Hence, random partitioning is applied whenever quick and load balanced assignments are required. A randomized approach to the minimum cut problem for weighted undirected graphs was presented by Karger *et al* [18]. Soule *et al* present the use of a round robin assignment which is a simple modification to the random partitioning algorithm [33].

3.2. Search Based Partitioning. In depth-first strategies, a depth-first traversal of the circuit graph is conducted [17]. The search begins from primary input nodes in the graph. Primary input nodes denote gates in the logic circuit that feed inputs to the circuit. The gates are assigned to partitions in the order visited. This strategy exploits the presence of linear chains of gates in the circuit and hence communication overheads are reduced to an extent. However, the issue of communication and concurrency is not addressed simultaneously. Breadth-first search partitioning schemes (also called the cluster partitioning scheme) are similar to depth-first strategy except that a breadth-first search is conducted. The breadth-first scheme clusters gates that are connected together in the same partition. Depending on the circuit model, the breadth-first scheme might reduce communication but places no emphasis on concurrency. In any case, load balance can be maintained in both strategies.

3.3. Topological/Level Partitioning. The techniques described above do not directly address the problem of concurrency. Unlike such techniques, the topological partitioning scheme [10, 31] attempts to improve concurrency by a level sorting scheme. In this strategy, the nodes in the circuit graph are first level sorted according to their topological level in the graph. Nodes at the same topological level are then assigned to different partitions. Load balancing does not pose a problem in this scheme. Concurrency is greatly improved but increased communication renders this scheme ineffective in many cases.

3.4. Partitioning by Element Strings. Prathima presents a technique for partitioning using element strings and gate delays [1]. This technique also proceeds from primary inputs to primary outputs in the graph. Initially, linear chains of gates starting from the input are assigned to partitions. A linear chain is a string of gates such that each gate has exactly one of its fanins and exactly one of its fanouts in the chain. This step encourages concurrency during simulation. In the next step, every assigned gate is examined for multiple fanouts. Every such unassigned gate on its fanout, along with its linear chain, is assigned to the same partition only if its gate delay is different from the delays of other gates (assigned to the same partition) on the same fanout. Due to this, scheduling of future events occur at different simulation times within a partition, thereby reducing communication. The second step might create load imbalance. Hence, a third step to move strings across partitions for load balancing is employed. This scheme exhibits good concurrency and communication features [1].

3.5. Fanout/Fanin Cone Partitioning. A partitioning scheme based on fan-in/fanout cone clustering starting from the input gates was developed and studied by Smith *et al* [31]. The main concern in this scheme is to reduce inter-processor communication. The first step in this scheme is to collect the set of fanout/fanin cones for every gate in the circuit graph. A fanout cone of any gate is the set of gates that are affected by the output of this gate. Similarly, a fanin cone of a gate is the set of gates that affect the inputs of the gate. After collecting the fanout/fanin cones, the primary input gates are distributed equally across partitions. An unassigned gate is then randomly selected and assigned to the partition with which it has maximum cone overlap. One common variation of this scheme is to allow replication of gates across partitions. An improvement to reduce replication in cone clustering through iterative refinement was reported by Manjikian [23].

3.6. Corolla Partitioning. Sporrer [35] proposed a new hierarchical partitioning approach to partition VLSI circuits for distributed logic simulators. The main idea behind this approach is to first identify strongly connected regions called petals and treat them as atomic (inseparable) components while partitioning. The algorithm proceeds in two well defined steps. In the first step, fine grained clusters with minimal interconnection are formed. The clusters are formed with no consideration for partition sizes. A second step combines clusters together in a partition in an attempt to load balance the partitions. The Corolla approach was demonstrated to yield good partitions for large circuits [35].

3.7. Concurrency Preserving Partitioning. A linear time partitioning algorithm that preserves concurrency based on instantaneous workload was developed by Hong [21]. Partitioning was achieved by placing gates that could be concurrently evaluated in different partitions. This algorithm consists of three phases where the three competing goals of communication, concurrency, and load balance are separately addressed. The first phase addresses concurrency by forming sub-graphs of primary input gates along with a sizable number of gates that can be reached from the primary input gate. The second phase attempts to topologically order gates within each sub-graph through a depth-first search starting from the input gates. Inter-processor communication is considered in the second phase. The third phase assigns gates at the same topological level to different partitions. While assigning gates to partitions, load balancing is achieved through a simple procedure using node activity levels. This heuristic was shown to produce reasonable speedup for several benchmarks [21].

3.8. Other Partitioning Strategies. Among other algorithms, Patil *et al* [27] developed several heuristics for partitioning, based on a cost function, for partitioning. The heuristics include greedy and simulated annealing techniques. The algorithm uses a cost function that estimates the execution time for parallel simulation given the processor assignment and the underlying architecture. Bagrodia *et al* [2] have illustrated the use of an acyclic multi-way partitioning scheme for gate level simulations. The technique uses a modified "k-way" generalization of the Fidduccia-Mattheyeses algorithm [11]. A parti-

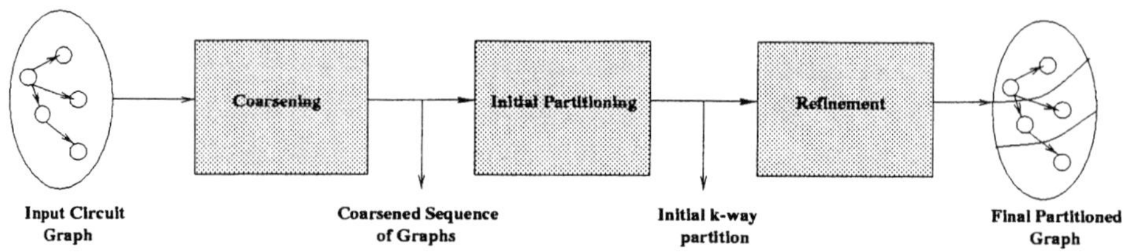

FIG. 4.1. *Overview of the Multilevel Partitioning Algorithm*

tioning scheme called "sensitive partitioning" based on propagation delays was illustrated by Pavlos *et al* [22] for synchronous parallel simulation [22]. An analysis of several approaches to partitioning for parallel logic simulation is available in literature [10, 31]. A number of other partitioning algorithms have been reported in the literature [4, 5, 7, 24, 32]. In this study, the following partitioning algorithms are compared to our multilevel approach: (i) fanout/fanin cone clustering, (ii) random partitioning, (iii) the depth-first search partitioner, and (iv) the topological partitioner.

4. The Multilevel Approach. This section will introduce the new partitioning technique we developed that is based on a multilevel approach. The multilevel approach to partitioning attempts to optimize the concurrency, communication, and load balance by decoupling them into separate phases. The multilevel algorithm for partitioning was studied and analyzed in [14, 19] and has been shown to produce high quality partitions (measured with respect to edges cut, *i.e.*, the number of edges that cross partition boundaries) over several partitioning algorithms such as the inertial and the spectral bisection algorithms [14, 19]. In addition, the complexity of the multilevel algorithm is $\mathcal{O}(\mathcal{N}_\mathcal{E})$, where N_E represents the number of edges in the circuit graph making the multilevel partitioning technique a fast linear time heuristic.

An ideal partitioning of a circuit graph implies that the load is perfectly balanced and an equal number of gates are active in each partition at any simulation instance. A particular partitioning algorithm cannot deliver ideal partitioning for all circuit graphs because each circuit has its own structure and pattern of communication. The multilevel algorithm attempts to satisfy such constraints by separating out these concerns in three phases, namely: (i) the coarsening phase, (ii) the initial partitioning phase, and (iii) the refinement phase. A simplified illustration of the multilevel approach is shown in Figure 4.1. The coarsening phase aggregates sets of vertices from the input circuit graph model and produces a hierarchical sequence of coarse-grained graphs with the goal of increasing concurrency. The initial partitioning phase assigns vertices of the coarsest (also hierarchically lowest) graph to partitions such that concurrency and load balance are maintained. The final refinement phase attempts to reduce communication and balance load by refining every coarse-grained graph starting from the coarsest level to finest/original circuit graph level. Unlike other algorithms, the strength of the multilevel approach stems from the fact that it allows

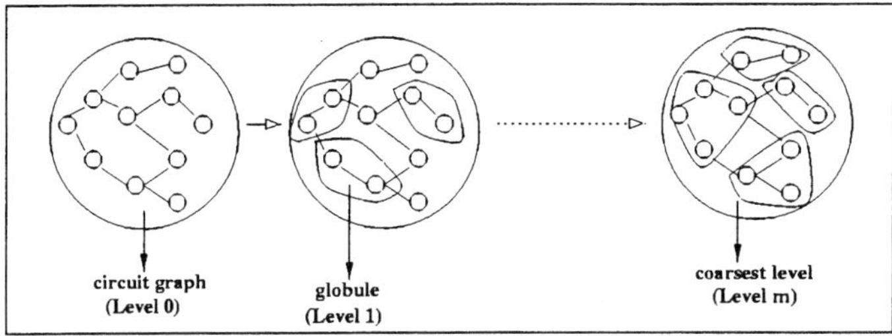

FIG. 4.2. *Coarsening procedure*

refining the circuit graph at several intermediate coarser levels instead of at only the original circuit graph level. This section presents a detailed description of the multilevel methodology for partitioning along with the necessary adaptations employed to fit it into the optimistic simulation framework. A detailed description of the three phases is presented in the following subsections.

Phase 1: Coarsening. Coarsening represents the first phase of the multilevel approach to partitioning. The chief goal of this phase is to improve concurrency by creating a hierarchical sequence of graphs derived from the original graph with smaller and smaller number of vertices. The various steps involved in the coarsening phase are shown in Figure 4.2. The coarsening phase proceeds from the primary input nodes in the graph. This is done in order to divide primary inputs across processors in the latter stages to encourage concurrency. The circuit graph of the initial set of processes (gates or nodes) constituting the circuit to be simulated is represented by G_0. The coarsening phase produces a hierarchical sequence of smaller graphs, say G_1, G_2, ..., G_m from the original graph G_0. Each intermediate graph G_i represents a different *view* of the original circuit graph. Each vertex (also called globule) in a lower level graph, say G_1, represents a set of connected vertices in its immediate higher level graph G_0. Each stage in this phase coarsens (or subsumes) a set of inter-connected vertices to yield a single vertex in the next stage. Emphasis on concurrency is stressed in this phase. In essence, distributing the objects in a concurrent manner reduces the number of *rollbacks* [13], that occur during optimistic simulations, and improves performance.

Coarsening of a circuit graph can be achieved using different coarsening schemes based on various parameters. Coarsening, irrespective of the scheme used, produces a sequence of smaller graphs derived from the original graph G_0. Any scheme simply defines the manner in which coarsening proceeds. As shown in Figure 4.2, coarsening starts from input vertices and combines a set of vertices from a higher level to form new vertices or globules in the next lower level. At each level, a vertex is allowed to be coarsened only once and vertices that contain a primary input vertex are not allowed to be combined together. This restriction is placed in order to maintain concurrency. The coarsening procedure halts when the number of globules fall below a threshold or if all the globules are input globules preventing further combination. The resulting graphs from coarsening satisfy the following relation. Given G_0 is the original graph, $G_i = (V_i, E_i)$, and $G_{i+1} = (V_{i+1}, E_{i+1})$ are graphs at levels i and $(i+1)$, it follows that:

$$V_{i+1} = \{V_{(i+1),0}, V_{(i+1),1}, ..., V_{(i+1),n}\}, \text{ and}$$
$$V_{(i+1),k} \subset V_i, \text{ and}$$
$$V_{(i+1),k} \cap V_{(i+1),l} = \phi.$$

The above equations define the relation between vertices at arbitrary levels i and $(i+1)$. In a nutshell, they imply that the vertices at level $(i+1)$ represent a group of vertices from the vertex set at level i. In addition, the edge set of a vertex at level $(i+1)$ then becomes the union of the edges of vertices at level i from which it was originally composed.

In the implementation reported herein a *fanout coarsening* scheme is employed in order to improve concurrency of parallel simulations. If vertices on interconnecting signals are split across partitions, graphs with a number of vertices on their signals/edges tend to increase the number of *rollbacks* [13] in optimistic Time Warp simulations. Fanout coarsening avoids this by maintaining vertices on a signal together in a partition; thereby reducing communication across partitions. In this technique, coarsening begins from primary input vertices and proceeds in a depth-first manner. Whenever a vertex is chosen for coarsening in this scheme, it is combined with all other vertices on its fanout. A vertex could be connected to several signals, however, only one of them is considered for coarsening. At each level, other than the first, coarsening starts from vertices that were just added to a globule in the previous level, thereby increasing concurrency with linear chains. To speed up coarsening and also reduce the number of levels, coarsening can be start at other random vertices. Other coarsening schemes could also be devised depending on the nature of communication, gate delays, and structure of the graph.

Phase 2: Initial Partitioning. The initial partitioning phase forms the second stage of the multilevel partitioning approach. Initial partitioning at the coarsest level provides a "k-way" partitioning of the original graph. The value of k is determined by the number of partitions desired. This phase attempts to load balance by distributing equal number of vertices across partitions while preserving concurrency. This phase is also the first phase responsible for assigning vertices to partitions. The partitions generated in this phase are further refined in the next phase of the multilevel algorithm. The initial partition before refinement is shown in Figure 4.3 using dotted lines.

Initially, all the input globules in the coarsest level are split equally across the partitions such that the load is sufficiently balanced. Any remaining globules are assigned to partitions in a random manner, maintaining load balance. The next phase refines the initial partitioning by moving globules across partitions to reduce communication. When a coarsest level globule is assigned to a partition,

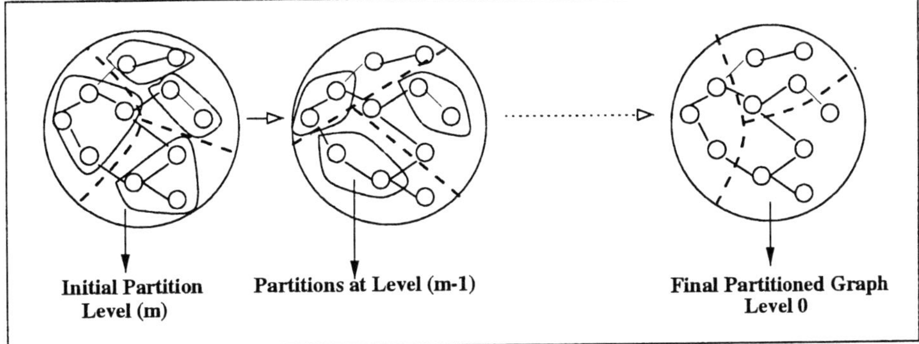

FIG. 4.3. *Refining procedure*

all vertices that make up this globule belong to this partition. Although this globule was formed via combination of globules over several levels (starting from the first), it actually represents a set of vertices from the original circuit graph. Hence, the main task of the initial partitioning phase is to map gates from the original circuit graph to partitions and produce an initial assignment. The following relation expresses the same fact; let the lowest level be m, V_{ij} be the jth vertex at level i ($0 \leq i \leq m$), and $P[V_{ij}]$ be partition to which vertex V_{ij} was assigned; it can be shown that:

$$\forall_{(v \in V_{ij})} P[v] = P[V_{ij}]$$

where v corresponds to a vertex from level $(i-1)$ that has been combined with other vertices from level $(i-1)$ to form V_{ij} at level i. At the end of initial partitioning, an initial load balanced assignment of gates is generated for refinement in the next phase.

Phase 3: Refinement. The coarsening and the initial partitioning phases concentrated on improving concurrency and load balance. The refinement phase, which corresponds to the third phase in this approach, attempts to reduce communication by placing strongly connected globules together in a partition. Beginning from the lowest coarse-grained level, this phase tries to load balance and reduce communication through "k-way" refinement at each intermediate level. The "k-way" partition of the original graph generated by the initial partitioning phase is utilized. The greedy algorithm for local refinement was used for the following three reasons:

(i) The greedy algorithm converges in a few iterations reducing the time needed for partitioning.

(ii) The greedy technique has also been shown to yield better partitions [19] with reduced edge-cut compared to other refinement algorithms (*e.g.*, Kernighan-Lin [20] and Fiduccia-Mattheyses [11]).

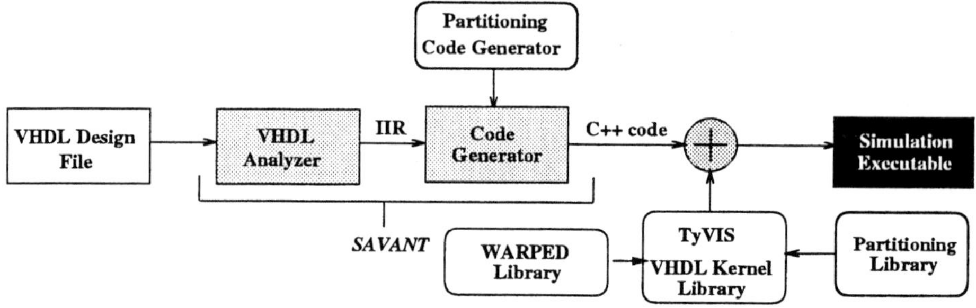

FIG. 5.1. *The Simulation Framework*

(iii) Along with the attempt to reduce communication, concurrency should also be maintained in this phase. If either the Kernighan-Lin [20] or the Fiduccia-Mattheyses [11] algorithm is employed, too many moves might result in a completely different configuration with reduced concurrency at the original circuit graph level. Since the greedy algorithm executes a relatively lower number of moves, concurrency is still maintained.

This phase starts from the lowest level of the hierarchical sequence of graphs, namely level m. The cut-set, that represents the number of edges that cross over partitions, is used as the parameter to be minimized by the greedy algorithm. The greedy refinement algorithm selects a vertex at random and computes the gain in the cut-set (reduction in edge-cut) for every partition that the vertex can be moved. The partition with maximum gain is then selected for the move. A move is feasible if it reduces the cut-set and preserves load balance. Reducing the cut-set in turn reduces communication overheads. Once a vertex is selected for a move, it is "locked", preventing its move until an iteration of the greedy algorithm finishes. The greedy algorithm will converge within a few iterations. After every move/iteration, a new assignment of vertices is generated. As illustrated in Figure 4.3, the graphs are recursively projected to the next higher level, preserving the assignment information, while refining across levels. The final assignment of gates to partitions is then generated at the original circuit graph level.

5. Experimental Framework. The simulation framework used for the partitioning studies is illustrated in Figure 5.1. The simulation framework was developed at the University of Cincinnati as a part of the SAVANT [36] project. As illustrated in Figure 5.1, it consists of three main components; SAVANT, TYVIS, and WARPED. The current implementation of the framework is in C++. A brief description of each component is provided in the following subsections.

5.1. SAVANT. The primary input to the framework is the description of the circuit in VHDL [15]. The VHDL analyzer (`scram`) is built using the Purdue Compiler Construction Tool Set (PCCTS) parser generator [26]. The input VHDL is analyzed into the standard intermediate form called Advanced

Intermediate Representation with Extensibility (AIRE) [37]. AIRE uses object oriented techniques to provide efficient access and ease of use of the intermediate form (IF). This representation is simply a translation of the VHDL model into a form that is easy to use, access, and extend for other tools at the front end. Hence, this intermediate form provides a base for sharing design information across various tool components. Furthermore, several application specific classes can be added to the hierarchy to extend the representation for application specific needs. For example, the IF is used by the code generator (Figure 5.1) to generate C++ code compliant with the TyVIS interface for VHDL simulation by interposing code generator specific classes and methods into the AIRE representation. The generated code is compiled and linked with the TyVIS and WARPED libraries to obtain the final simulation executable. A library manager was also built in to create and maintain design libraries which maybe referenced by other VHDL descriptions.

5.2. TyVIS. TyVIS is a VHDL kernel that provides the necessary runtime support for simulation of VHDL designs. It provides the basic data structures and methods necessary to interface the C++ code generated by `scram` with the WARPED [28] parallel simulation kernel. TyVIS operates on top of WARPED which provides the parallel discrete event simulation capability. However, the operation of the WARPED simulation kernel is completely transparent to the TyVIS kernel. The routines needed for partitioning were integrated along with the code-generator and the TyVIS kernel. Model specific data that was needed for partitioning was extracted at code-generation time. The partitioning methods that get invoked during simulation are also generated. These routines provide support to build suitable data structures for partitioning. Object oriented techniques have been exploited in the implementation of the runtime support routines and the data structures used for partitioning. This generated code provides sufficient flexibility to implement several partitioning algorithms without any changes to `scram`.

5.2.1. The Partitioning Framework. The necessary infrastructure to enable partitioning of the VHDL processes was integrated with the TyVIS kernel. This was achieved by developing a generic framework for partitioning and integrating it with the TyVIS kernel. The main design issues that are immediately evident are (i) the framework must ease the implementation of the different partitioning algorithms, and (ii) other partitioning algorithms can be simply plugged in if they comply to some interface specification thereby facilitating future extensions. With the above issues in mind, the *application program interface* (API) (illustrated in Figure 5.2) for partitioning was designed. The object oriented design consists of a base class, called `PartitionBase`, that contains information common to all partition algorithms. Partitioning algorithms are then derived from `PartitionBase` and implemented. The interface specification for the functions provided in Figure 5.2 are:

(i) `void initialize()`: This method is invoked just before elaboration [36] of the input VHDL design to create and initialize the data structures

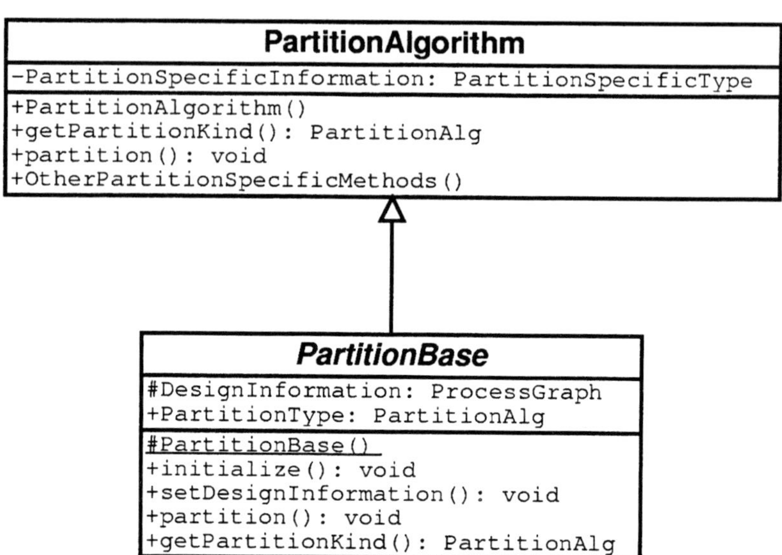

FIG. 5.2. *The API for the Partitioning Framework*

needed for partitioning. This method is defined in the base class and should not be overloaded by the child classes.

(ii) **`void setDesignInformation()`**: This method provides access to filling the data structures (**`DesignInformation`**) that are necessary while partitioning. It is actually a set of functions, but for the sake of clarity, presented here as a single method. While the design is being elaborated, the data structures are gradually filled in with design information. This method, again should not be overloaded by the child classes.

(iii) **`void partition()`**: This method must be overloaded in the derived classes. According to the scheme required for partitioning, appropriate code is embedded into this method to implement the functionality. The derived classes (**`PartitionAlgorithm`**) might include data structures and functions specific to the partitioning scheme in **`PartitionSpecificInformation`** and **`OtherSpecificMethods()`** respectively.

(iv) **`PartitionAlg getPartitionKind()`**: This method identifies the partitioning algorithm currently being used and is useful for debugging purposes.

A set of six different partitions strategies, including the multilevel strategy, were derived from **`PartitionBase`** and implemented. The partitioning algorithms used the processes generated at the end of the elaboration phase. Since the framework employs a runtime elaboration technique [36], partitioning occurs at runtime, after the simulation objects/processes are instantiated. The runtime support functions generated by **`scram`** are used by the partitioning algorithms to build the necessary data structures. After elaboration, the partition algorithm instantiated then invokes the corresponding **`partition()`** routine which computes the assignment according to the number of partitions desired. Hence, a runtime technique to choose the number of partitions was incorporated. In addition, the runtime partitioning technique was designed to provide the flexibility to choose from different partitioning algorithms without requiring a complete recompilation of the system. Once the design has been partitioned, the simulation is initialized and executed. The partitioning framework provides an efficient and robust subsystem for partitioning studies.

5.3. WARPED. WARPED is an optimistic parallel discrete event simulator developed at the University of Cincinnati. It uses the Time Warp mechanism [13, 16] for distributed synchronization. In WARPED, the logical processes (LPs) that represent the physical processes being modeled are placed into groups called "clusters". The clusters represent the operating system level parallel processes constituting the simulation. LPs on the same cluster communicate directly with each other without the intervention of the messaging sub-system. This technique enables sharing of events between LPs which not only reduces the communication overheads but also reduces memory overheads in the system. Communication between the clusters is achieved using one of several possible message passing systems (*e.g.*, MPI, MPI-BIP, TCP) [30]. LPs within a cluster operate as classical Time Warp processes. WARPED presents a simple and robust object-oriented application program interface (API) for model development. Further details on the working of TyVIS and WARPED are available elsewhere [36].

Several optimizations to improve the performance of the WARPED kernel have been implemented [9, 12, 25, 29]. The *message aggregation* strategy was one such optimization implemented in the WARPED kernel that was employed to reduce the communication overhead in parallel simulations [9, 8, 30]. Message aggregation is the process of collecting messages in close temporal and spatial proximity and delivering them as a single physical message. For example, messages are sent to the same destination are spatially proximal. Consecutive messages at similar wall clock times exhibit temporal proximity. Several strategies were devised to study the effect of message aggregation on performance [9, 8]. The performance results presented here includes studies conducted with the *simple adaptive aggregation window* (SAAW) strategy for message aggregation [8].

6. Benchmarks. Three of the ISCAS '89 benchmarks [6] were used to evaluate the performance of the partitioning algorithms. The characteristics of the models used in the experiments are shown in Table 6.1. The ISCAS '89 benchmarks were primary designed to extend the size and complexity of the ISCAS '85 [34] benchmarks. All the ISCAS '89 benchmark circuits are sequential in nature and use D-type flip flops, while the ISCAS '85 benchmarks are combinatorial. These benchmarks were originally generated to evaluate full scan-based algorithms for test pattern generation and identify limitations of these algorithms. The functional descriptions of the test circuits are not available in literature. The benchmarks presented in Table 6.1 are real chip based and rely on partial scan. The size of the benchmarks used in the experiments was limited due to the limitations on the number of resources available for simulation. The

TABLE 6.1
Characteristics of benchmarks

Circuit	Inputs	Gates	Outputs	Interconnect
s5378	35	2779	49	2909
s9234-1	36	5597	39	5769
s15850-1	77	10383	150	10156

limitation was partly imposed due to the Time Warp algorithm that consumed lots of memory for state saving and hence resulted in thrashing. In fact for larger and larger circuits, the number of states saved increases with the number of gates and simulation execution time. Therefore too much of swapping overwhelmed the simulation process. In some cases, we also ran out of virtual memory. It must be noted that the experiments were conducted with the base version of Time Warp and did not include several optimizations. The optimizations were turned off in order fully isolate and study the effects of partitioning. For example, we did not experiment with several state saving mechanisms (we saved every state, default state saving) since it was not the focus of study in this paper. Furthermore, we have not experimented with other synchronization strategies (such as conservative synchronization strategies) since the focus of this study is on partitioning. Hence, it is impossible even to hypothesize about the effectiveness of other synchronization strategies in this study. However, the effectiveness of the multilevel partitioning algorithm can be estimated using analytical techniques. The time complexity of the multilevel partitioning technique is $O(E)$, where E is the number of edges/signals in the given circuit graph/circuit. Accordingly, the performance of the algorithm is linear with respect to the number of edges. Our experiments the multilevel algorithm conforms to the time complexity estimates and appears to scale well. Hence, we believe that the algorithm can be effectively applied (with suitable parameters) to large circuits consisting of millions of gates.

The partitioning techniques used in the experiments include the multilevel methodology and the following five algorithms. A brief description of the algorithms is given below.

1. **Random Partitioning**: The gates in the input circuit graph are assigned to partitions in a random manner. Load balance is maintained by distributing approximately equal number of gates into partitions. Random partitioning does not guarantee reduction in communication or improvement in concurrency.

2. **Cluster Partitioning**: A bread-first traversal of the circuit graph, beginning from the inputs, is used to assign gates to partitions in the order visited. This technique clusters processes keeping gates that are connected, in the same partition. Load balance is again maintained as in the case of random partitioning. Depending on the model, this scheme might reduce the amount of communication but places no emphasis on concurrency.

3. **Topological Partitioning**: In this scheme, the circuit graph is first level-sorted and then gates with consecutive numbers are assigned to partitions.

Load balance is maintained, concurrency is greatly improved but increased communication renders this scheme ineffective in most cases.

4. **DFS Partitioning**: This scheme is similar to the clustering technique except that a depth-first is conducted. This scheme reduces communication to an extent but the issue of communication and concurrency together, is not addressed.

5. **Fanout/Fanin Cone Partitioning**: This technique is based on fanout-fanin cones [31]. Initially, all the input gates are divided across partitions equally. A vertex chosen for assignment is then placed in a partition with which it has maximum fanout/fanin cone overlap. Reduction in communication is achieved since connected gates are placed together. However, this technique does not address issues related to concurrency. Fanout cone partitioning was used.

7. Experiments, Results and Analysis. The parallel simulation experiments were conducted on eight workstations inter-connected by fast Ethernet. Each workstation consisted of dual 300MHz Pentium II processors with 128MB of RAM running Linux 2.2.12. The MPICH implementation of the MPI interface for message passing was employed for communication between workstations. The experiments were conducted in the SAVANT/TyVIS/WARPED simulation framework presented in §5. The results presented in this section are for the s9234 benchmark. Studies conducted with message aggregation are also presented. The factors that were used to compare the different schemes are:

(i) **Simulation Execution Time**: This factor measures the time taken by the simulation to complete without the overhead of partitioning and presents an overall idea of the performance of a particular partitioning algorithm.

(ii) **Communication**: To study the amount of interprocess communication, the number of messages sent across partitions is collected for each simulation run and analyzed.

(iii) **Rollbacks**: This parameter represents the degree of concurrency present in the simulator. If the number of rollbacks are reduced, the simulation progresses at a more even rate.

(iv) **Workload**: In order to compute the workload of an entire simulation, the number of messages that are processed by all the logical processes is collected.

(v) **Load Balance**: The standard deviation from the mean workload of each processor is measured and used as the measure for load balance.

(vi) **Speedup**: Speedup is defined as the ratio of the time taken to simulate a model in parallel to that in sequential. Using the parallel simulation execution times and the sequential simulation execution times, this factor is calculated.

(vii) **Complexity**: This factor represents the time needed to partition a particular circuit.

7.1. Simulation Execution Time. The simulation times for the s9234 benchmark are shown in Figures 7.1 and 7.2. It is observed that the multilevel algorithm outperforms most of the partitioning algorithms when more than 4 workstations are involved in the simulation with or without message aggrega-

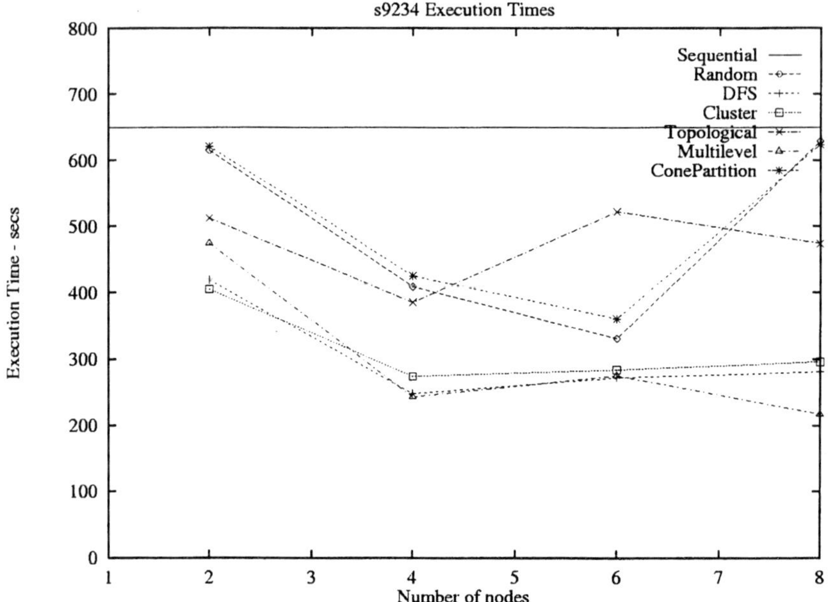

FIG. 7.1. *Execution times of s9234 without Message aggregation*

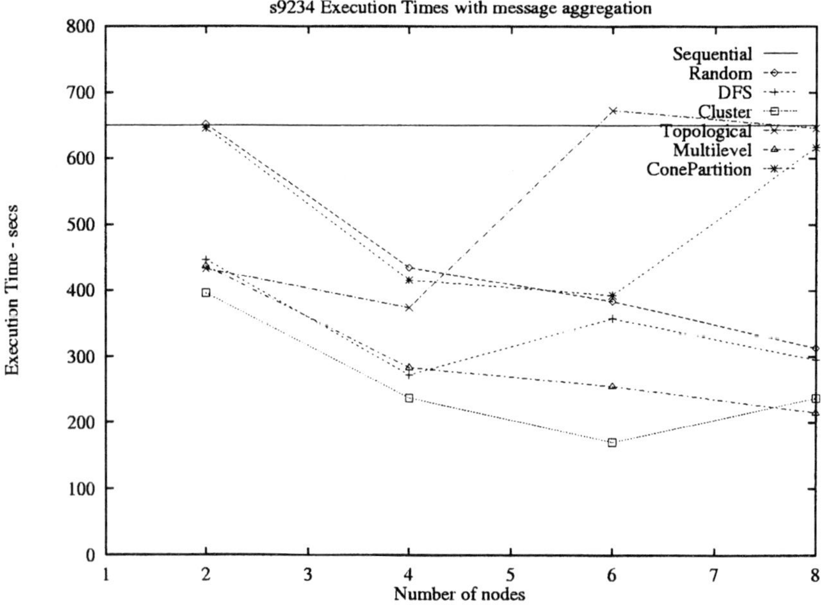

FIG. 7.2. *Execution times of s9234 with Message aggregation*

tion. The performance of the cluster and the depth-first algorithms deteriorates with increase in the number of nodes due to lack of concurrency, when message aggregation is not used. This lack of concurrency also increases the number of rollbacks in the simulations. Since the signals are split across partitions for concurrency, the performance of the Topological algorithm is limited due increased communication overheads. The simulation execution times for all the benchmarks (without message aggregation) have been tabulated in Table 7.1. When message aggregation is employed, the cluster partitioning algorithm begins performing better than the multilevel strategy. Aggregation across partition boundaries helps to reduce the communication overhead thereby lowering the execution times. Topological partitioning with the message aggregation performs even worse since message aggregation seems to inhibit the progress of the simulation. Since the cutset is greater, a lot of messages are aggregated and hence held back, inhibiting the progress of the simulation. The performance of the depth-first scheme is not affected very much. Similarly, the message aggregation simply serves to smoothen out the curve for the multilevel strategy in the 6 LP case.

The simulation execution time results for the rest of the benchmarks without message aggregation are tabulated in Table 7.1. Execution times for the s15850 benchmark with 2 workstations are not shown in the Table 7.1 because the simulation exhausted memory and did not complete successfully. We do not include the overhead incurred for partitioning into the simulation timings because all schemes described are static schemes *i.e.*, they do not play any role during simulation. The partitioning schemes are executed once for every circuit (before simulation) and used henceforth for every simulation run. The partitioning overheads for the various partitioning algorithms are discussed in subsection 7.7. From the data, it is evident that the multilevel strategy provides excellent performance when additional processors are involved in simulation. When message aggregation is also employed, some of the partitioning schemes reacted with very poor/excellent performance. It is clear from this observation that two optimizations (message aggregation and partitioning), which provide increased performance when employed individually, do not necessarily result in a performance boost when applied together.

7.2. Communication Characteristics. The messaging characteristics for the simulation experiments conducted with the s9234 benchmark are presented in Figures 7.3 and 7.4. The data also includes the anti-messages because they are an integral part of the simulation/communication overheads. As shown in the figures, the multilevel algorithm reduces the amount of communication in the 8 to 16 processor region. The cone partitioner performed well due to lower communication and better concurrency features. Increased communication overheads due to greater edge cut in the case of the topological partitioner resulted in increased execution times. The random and the cone partitioning did not perform well in the 16 processor case due to similar reasons. Hence, it is evident that increased communication across partitions hinders the progress of

TABLE 7.1
Simulation Time (in secs) for the different partitioning algorithms without message aggregation

Circuit	Seq Time	No. of Nodes	Random	DFS	Cluster	Top.	Multi.	Cone
s5378	150	2	152	154	140	112	132	150
		4	102	80	126	234	101	113
		6	161	103	118	236	92	124
		8	136	142	126	433	77	110
s9234-1	651	2	615	419	405	512	474	621
		4	409	248	274	385	243	425
		6	331	272	284	522	275	361
		8	629	281	296	474	217	624
s15850-1	2154	4	1324	1043	918	1333	1214	1345
		6	989	1943	1012	1522	985	1108
		8	942	1179	1003	2688	734	2211

a simulation which in turn affects concurrency in the simulator. Message aggregation usually tends to reduce the amount of messaging across partitions. For example, the depth-first scheme exhibits reduced communication characteristics with aggregation. In most cases, the communication either remains the same or is reduced with aggregation. Depending on the algorithm used for partitioning, the performance could be affected in both directions, since a plethora of parameters depend on the amount of communication.

7.3. Rollback Characteristics. The rollback characteristics of the s9234 model are shown in Figures 7.5 and 7.6. As illustrated by the bar chart, the multilevel algorithm greatly reduces the number of rollbacks during simulation; highlighting the equilibrium achieved between concurrency and communication. The behavior of most of the partitioning algorithms can be explained with the help of both the rollback and messaging characteristics. The sudden dips in the execution time graph (Figure 7.1) are caused by a lower number of rollbacks and lower communication for most of the partitioning algorithms. For example, the cluster, depth-first and the topological algorithms suffered from a number of rollbacks due to more communication across partitions degrading their performance in the 16 processor case. Message aggregation increased the number of rollbacks for the cluster partitioning scheme in the 16 processor case thereby degrading its performance.

7.4. Load Balance Characteristics. The load balance characteristics for the s9234 benchmark are illustrated in Figures 7.7 and Figure 7.8. The load balance is measured as the standard deviation from the mean of the individual workloads of the processors involved. The workload for a particular processor is the total number of messages processed (both positive and negative messages) during simulation. From Figure 7.7 it is evident that all the algorithms experience a deviation from the mean workload. A reduced deviation shows a better

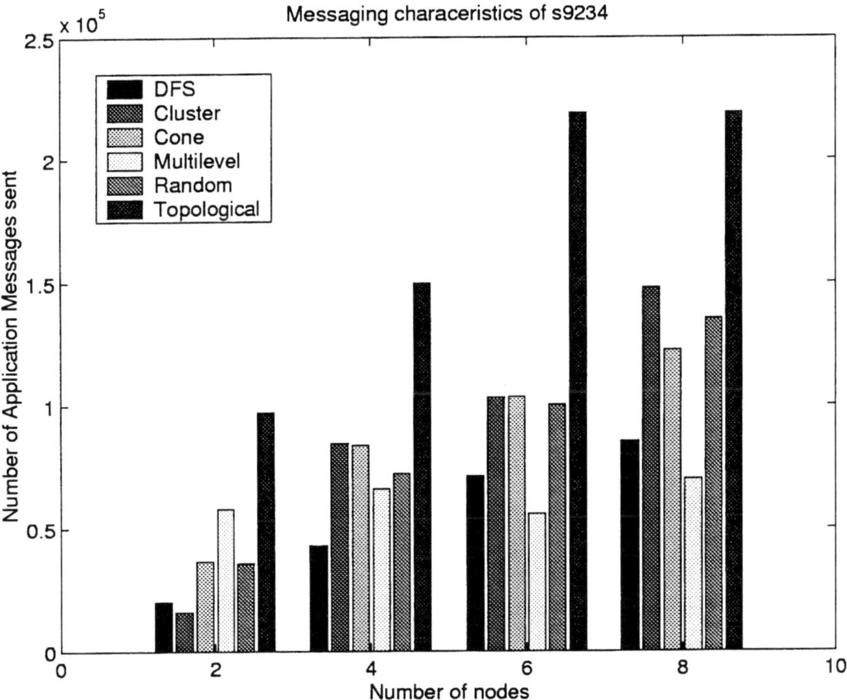

FIG. 7.3. *Messaging characteristics of s9234 without Message aggregation*

load balance. The multilevel algorithm does fall into the lower half of the deviations and hence provides decent load balance. It does not provide the best balance, since balancing load is not the only motive of the algorithm. Message aggregation resulted in increased deviations in most of the schemes, since the messages were aggregated and were not sent immediately to the destination LP's for processing. We can conclude on the basis of the bar charts that using message aggregation will have a negative effect on load balancing.

7.5. Workload Characteristics. The total number of messages processed by all the processors involved in the simulation was measured as the workload for the simulation. The workload characteristics for the s9234 benchmark are shown in Figures 7.9 and 7.10. Again, the multilevel algorithm does not give the best performance results for this characteristic. As explained earlier, optimizing several factors together resulted in tradeoffs between the various parameters. Furthermore, message aggregation tends to reduce the amount of workload processed.

7.6. Speedup. The speedup curves for the s9234 benchmark are illustrated in Figure 7.11 and Figure 7.12. The baseline used for computing the speedup is the execution time of the sequential simulator, available as a part of the WARPED

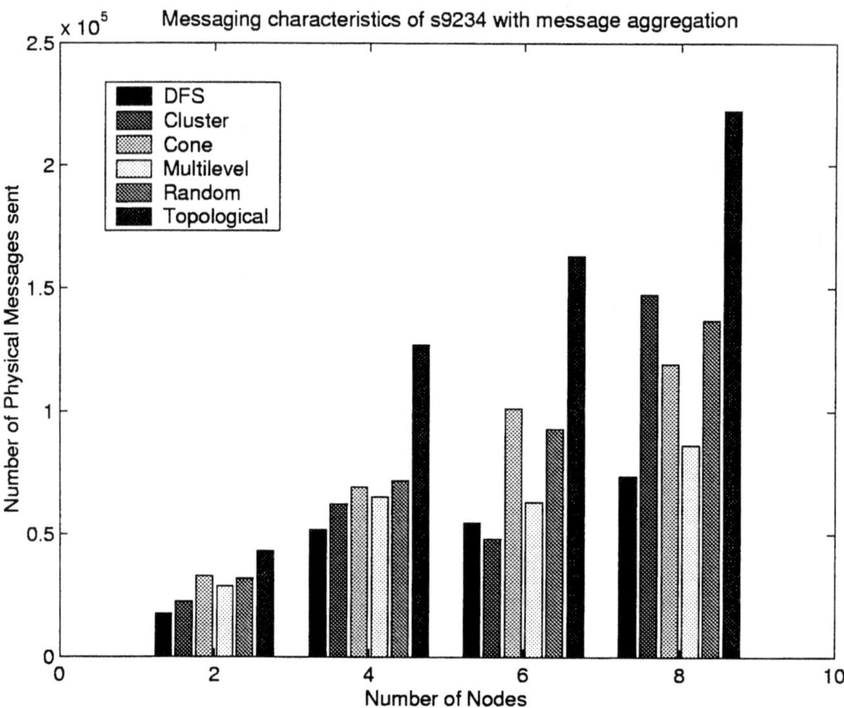

FIG. 7.4. *Messaging characteristics of s9234 with Message aggregation*

kernel. As shown in the Figure 7.12, the performance of the multilevel algorithm increases with the number of processors, indicating their scalability. The reduction in simulation time is due the availability of the increased computational resources that the partitioning algorithms effectively utilize. The algorithms extract more concurrency from the models as the available parallelism increases. The increase in computational resources also improves the speedup of some of the other algorithms when using 16 processors. The partitioning overheads have not been included in the speedup curves for two reasons. The primary reason being that, partitioning for a given circuit and for a given set of processors is done once and the generated partition is used in future simulation runs; thereby considerably minimizing (or even eliminating) the overall overheads of partitioning. The second reason being that, all the partitioning techniques are *static*, *i.e.*, partitioning occurs prior to simulation and the algorithms do not contribute any additional overheads during simulation.

It must also be noted that the experiments were conducted with the base version of Time Warp and did not include several optimizations. The optimizations were turned off in order fully isolate and study the effects of partitioning. For example, we did not experiment with several state saving mechanisms (such as periodic state saving, infrequent state saving), rollback reducing/eliminating

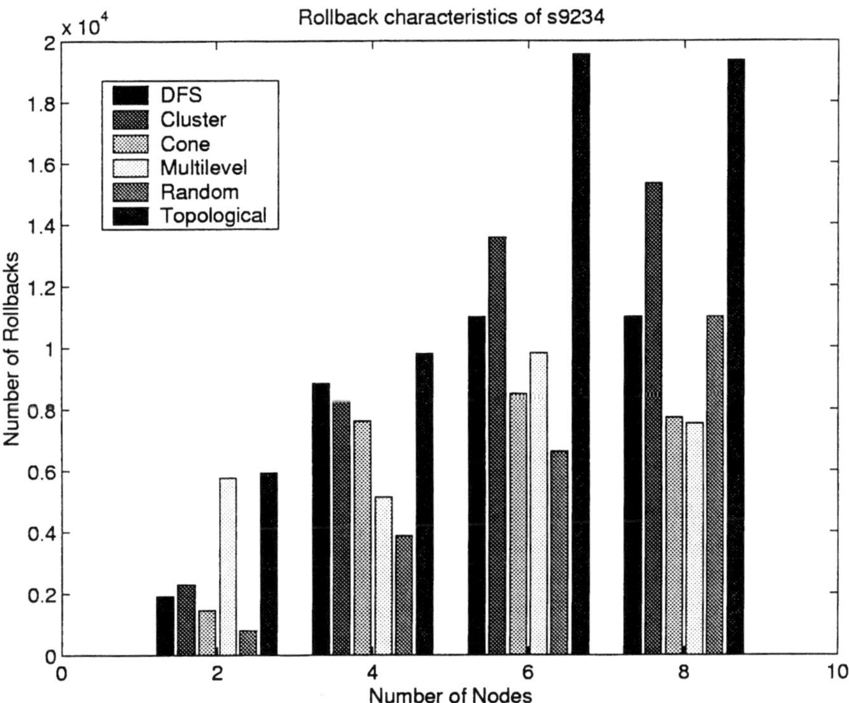

FIG. 7.5. *Rollback characteristics of s9234 without Message aggregation*

optimizations (such as rollback relaxation, lazy cancellation, dynamic cancellation, etc.) since it was not the focus of study in this paper. Furthermore, we have not experimented with other synchronization strategies (such as conservative synchronization strategies) since the focus of this study is on partitioning. Hence, it is impossible even to hypothesize about the effectiveness of other synchronization strategies, when compared to Time Warp, this study. Since the simulation kernel was configured for its most basic configuration, the overall improvement in performance is limited but the performance figures fully highlight the effectiveness of the various partitioning techniques.

7.7. Complexity. The complexity of any partitioning scheme represents the time needed by that scheme to compute an assignment for a circuit to a set of processors. The simpler schemes such as random, depth-first, cluster and topological partitioning schemes exhibit $O(n)$ complexity, where n represents the number of gates in the circuit. In contrast, the cone and multilevel schemes decide on assignments based on some heuristics. The complexity of the cone partitioner is $O(n^2)$ since it assigns gates to partitions based on already existing nodes in the partition and is not a linear time heuristic. The multilevel algorithm, on the other hand, is a linear time heuristic on the number of edges in the circuit, that is, $O(E)$ where E is the number of edges in the process graph.

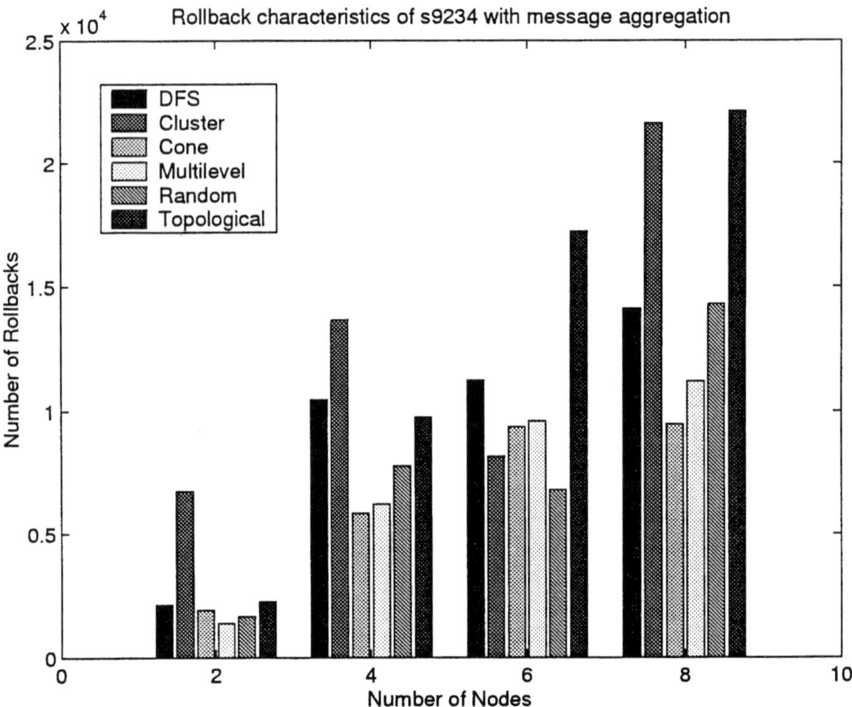

FIG. 7.6. *Rollback characteristics of s9234 with Message aggregation*

The partitioning times for all the algorithms is tabulated in Table 7.2.

8. Conclusions. In this paper, we presented the partitioning studies that were conducted on a parallel VHDL simulation framework (SAVANT/TYVIS/WARPED). A new partitioning technique based on the multilevel heuristic was developed and its performance relative to existing partitioning strategies was investigated. In addition, the issues involved in developing a generic framework for partitioning were discussed. The integration of the various partitioning strategies into the simulation framework through the partitioning framework was also described. The experiments conducted included message aggregation as one of the optimizations along with partitioning. Results from the experimental analysis indicate that the multilevel technique yielded better partitions than other partitioning strategies in most cases with or without message aggregation. Parallel simulation (of all the sample applications) on 16 processors using the multilevel technique executed in less than half the time taken by a sequential simulation of the same application(s). This speedup can be attributed to the reduction in both the number of rollbacks and the amount of inter-processor communication. With message aggregation applied as an optimization alongside partitioning, some of the partitioning schemes reacted with very poor/excellent performance. It is clear from this observation that two optimizations, which

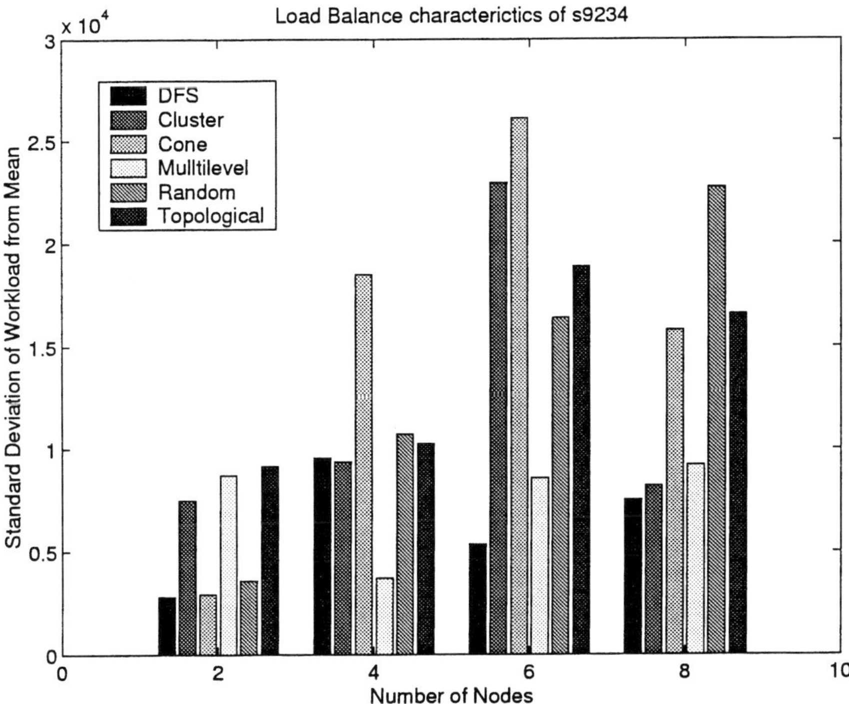

FIG. 7.7. *Load balance characteristics of s9234 without Message aggregation*

provide increased performance when employed individually, need not necessarily result in a performance boost when applied together. Since the multilevel technique is a linear time heuristic, it can be easily scaled to partition for a large number of processors. Research is currently ongoing to incorporate several enhancements to the multilevel heuristic. For example, we are currently investigating the use of activity levels of communication to make better decisions while coarsening. In addition, different schemes for coarsening and refinement are also being studied.

Acknowledgments. The authors would like to acknowledge the suggestions and contributions of Radharamanan Radhakrishnan and Malolan Chetlur to this work.

REFERENCES

[1] P. AGRAWAL, *Concurrency and communication in hardware simulators*, IEEE Transactions on Computer-Aided Design, (1986).

[2] R. L. BAGRODIA AND W. LIAO, *Maisie: A language for the design of efficient discrete-event simulations*, IEEE Transactions on Software Engineering, 20 (1994), pp. 225–238.

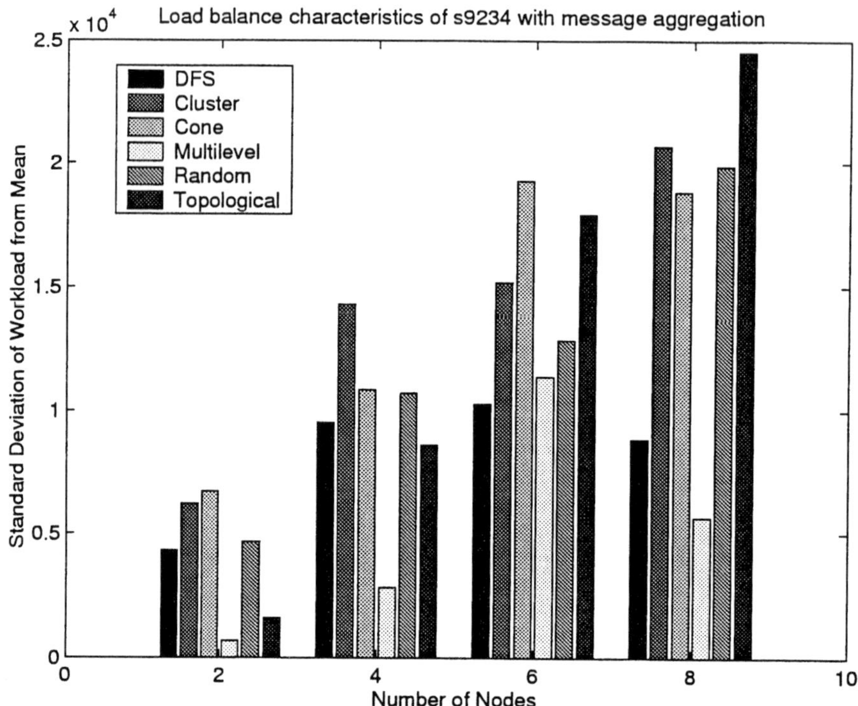

FIG. 7.8. *Load balance characteristics of s9234 with Message aggregation*

[3] M. L. BAILEY, J. V. BRINER, JR., AND R. D. CHAMBERLAIN, *Parallel logic simulation of VLSI systems*, ACM Computing Surveys, 26 (1994), pp. 255–294.

[4] A. BOUKERCHE AND C. TROPPER, *A static partitioning and mapping algorithm for conservative parallel simulations*, in Proceedings of the 8th Workshop on Parallel and Distributed Simulation (PADS'94), D. K. Arvind, R. Bagrodia, and J. Y.-B. Lin, eds., Baltimore, MY, July 1994, ACM Press, pp. 164–172.

[5] A. BOUKERCHE AND C. TROPPER, *SGTNE: Semi-global time of the next event algorithm*, in Proceedings of the 9th Workshop on Parallel and Distributed Simulation (PADS '95), 1995, pp. 68–77.

[6] CAD BENCHMARKING LAB , NCSU, *ISCAS'89 Benchmark Information.* (available at `http://www.cbl.ncsu.edu/www/CBL_Docs/iscas89.html`).

[7] R. D. CHAMBERLAIN AND C. D. HENDERSON, *Evaluation the use of pre-simulation in VLSI circuit partitioning*, in Proceedings of the 8th Workshop on Parallel and Distributed Simulation (PADS '94), 1994.

[8] M. CHETLUR, *Reducing communication overheads in asynchronous distributed applications*, master's thesis, University of Cincinnati, Dec. 1998.

[9] M. CHETLUR, N. ABU-GHAZALEH, R. RADHAKRISHNAN, AND P. A. WILSEY, *Optimizing communication in Time-Warp simulators*, in 12th Workshop on Parallel and Distributed Simulation, Society for Computer Simulation, May 1998, pp. 64–71.

[10] J. CLOUTIER, E. CERNY, AND F. GUERTIN, *Model partitioning and the performance of distributed time warp simulation of logic circuits*, in Simulation Practice and Theory, 1997, pp. 83–99.

[11] C. M. FIDUCCIA AND R. M. MATTHEYSES, *A linear time heuristic for improving network partitions.*, in Proceedings of the 19th IEEE Design Automation Conference, 1982,

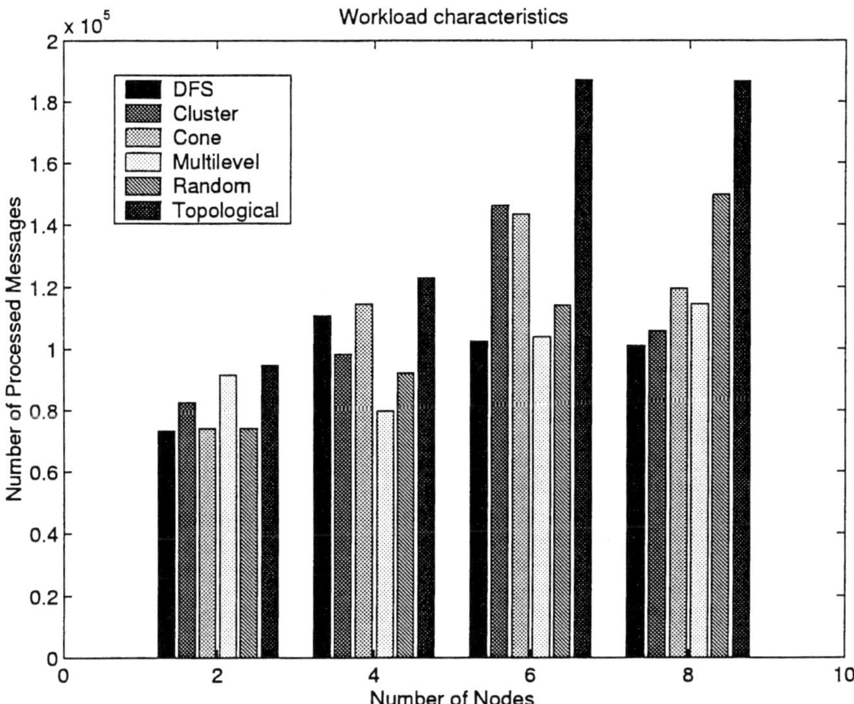

FIG. 7.9. *Workload characteristics of s9234 without Message aggregation*

pp. 175–181.

[12] J. FLEISCHMANN AND P. A. WILSEY, *Comparative analysis of periodic state saving techniques in Time Warp simulators*, in Proc. of the 9th Workshop on Parallel and Distributed Simulation (PADS 95), June 1995, pp. 50–58.

[13] R. FUJIMOTO, *Parallel discrete event simulation*, Communications of the ACM, 33 (1990), pp. 30–53.

[14] B. HENDRICKSON AND R. LELAND, *A multi-level algorithm for partitioning graphs*, in Proceedings of the 1995 ACM/IEEE Supercomputing Conference, Dec. 1995.

[15] *IEEE Standard VHDL Language Reference Manual*, New York, NY, 1993.

[16] D. JEFFERSON, *Virtual time*, ACM Transactions on Programming Languages and Systems, 7 (1985), pp. 405–425.

[17] K. L. KAPP, T. C. HARTRUM, AND T. S. WAILES, *An improved cost function for static partitioning of parallel circuit simulations using a conservative synchronization protocol*, in Proceedings of the 9th Workshop on Parallel and Distributed Simulation (PADS '95), 1995, pp. 78–85.

[18] D. KARGER AND C. STEIN, *A new approach to the minimum cut problem*, Journal of the ACM, (1996), pp. 601–640.

[19] G. KARYPIS AND V. KUMAR, *Multilevel k-way partitioning scheme for irregular graphs*, Technical Report TR 95-055, University of Minnesota, Computer Science Department, Minneapolis, MN 55414, Aug. 1995.

[20] B. W. KERNIGHAN AND S. LIN, *An efficient heuristic procedure for partitioning graphs*, The Bell Systems Technical Journal, (1970), pp. 291–307.

[21] H. K. KIM AND J. JEAN, *Concurrency preserving partitioning (CPP) for parallel logic simulation*, in Proceedings of the Tenth Workshop on Parallel and Distributed Sim-

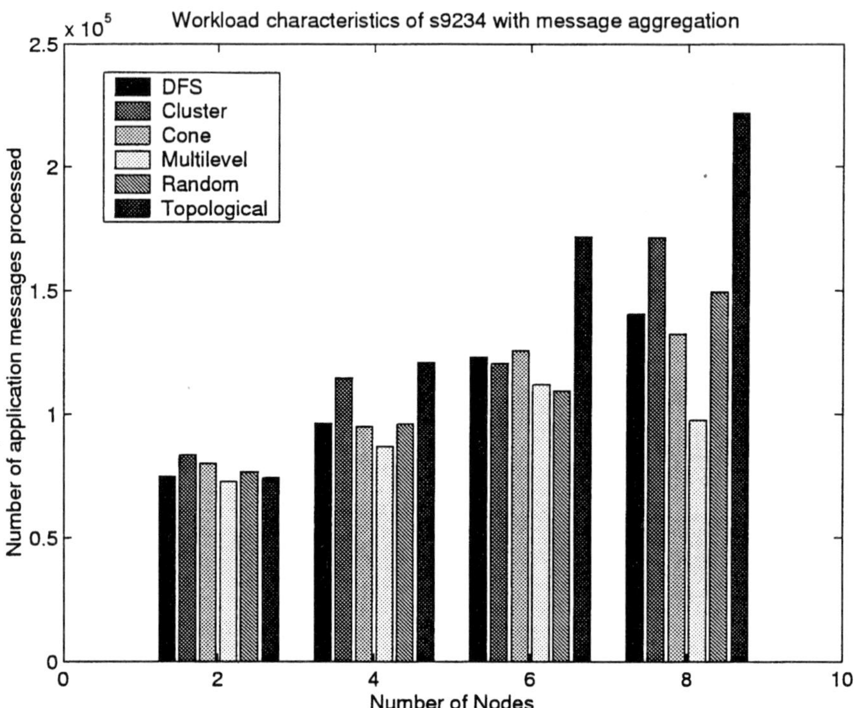

FIG. 7.10. *Workload characteristics of s9234 with Message aggregation*

ulation, May 22–24 1996, pp. 98–105.

[22] P. KONAS AND P. YEW, *Partitioning for synchronous parallel simulation*, in Proceedings of the 9th Workshop on Parallel and Distributed Simulation (PADS '95), 1995, pp. 181–184.

[23] N. MANJIKIAN AND W. M. LOUCKS, *High performance parallel logic simulation on a network of workstations*, in Proceedings of the 7th Workshop on Parallel and Distributed Simulation, May 1993, pp. 76–84.

[24] B. NANDY AND W. M. LOUCKS, *An algorithm for partitioning and mapping conservative parallel simulation onto multicomputers*, in Proceedings of the 6th Workshop on Parallel and Distributed Simulation (PADS '92), 1992, pp. 139–147.

[25] A. PALANISWAMY, S. AJI, AND P. A. WILSEY, *An efficient implementation of lazy reevaluation*, in Proc. of the 25th Annual Simulation Symposium, Society for Computer Simulation, Apr. 1992, pp. 140–146.

[26] T. J. PARR, *Language Translation Using PCCTS and C++*, Automata Publishing Company, January 1997.

[27] S. PATIL, P. BANERJEE, AND C. D. POLYCHRONOPOULOS, *Efficient circuit partitioning algorithms for parallel logic simulation*, in Proceedings, Supercomputing '89, Nov. 1989, pp. 361–370.

[28] R. RADHAKRISHNAN, D. E. MARTIN, M. CHETLUR, D. M. RAO, AND P. A. WILSEY, *An Object-Oriented Time Warp Simulation Kernel*, in Proceedings of the International Symposium on Computing in Object-Oriented Parallel Environments (ISCOPE'98), D. Caromel, R. R. Oldehoeft, and M. Tholburn, eds., vol. LNCS 1505, Springer-Verlag, Dec. 1998, pp. 13–23.

[29] R. RAJAN AND P. A. WILSEY, *Dynamically switching between lazy and aggressive can-*

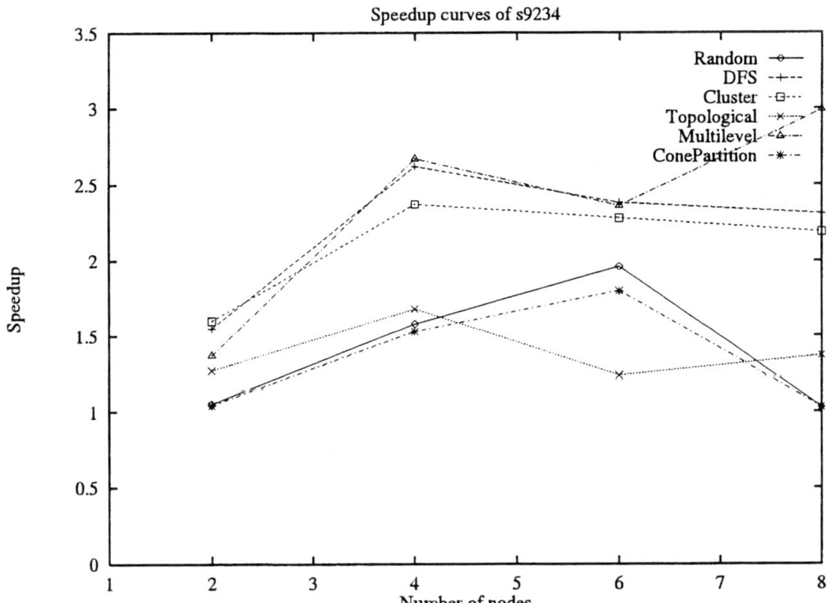

FIG. 7.11. *Speedup curves of s9234 without Message aggregation*

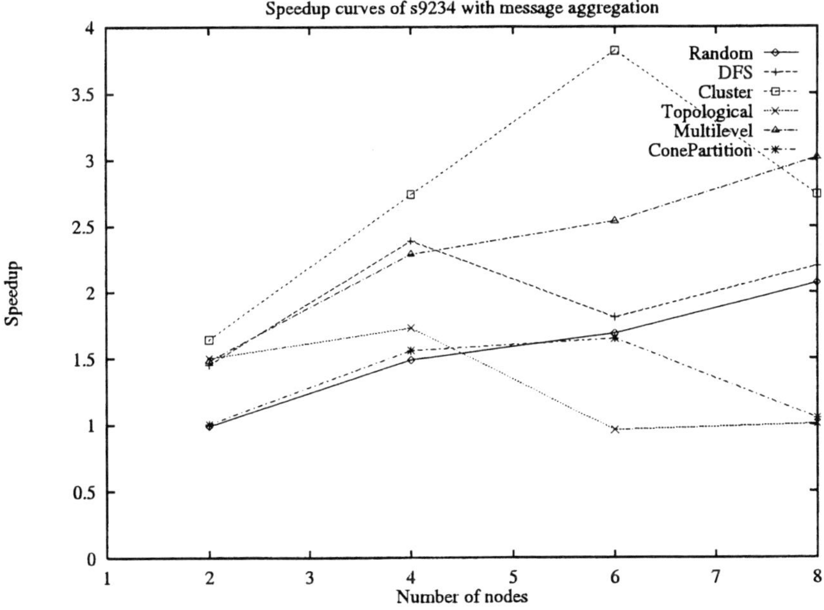

FIG. 7.12. *Speedup curves of s9234 with Message aggregation*

TABLE 7.2
Partitioning Times (in secs) for the different partitioning algorithms

Circuit	# Part.	Random	DFS	Cluster	Top.	Multi.	Cone
s5378	2	0.0003	0.0166	0.0346	0.0246	4.693	36.73
	4	0.0003	0.0167	0.0343	0.0245	11.907	36.07
	6	0.0003	0.0164	0.0346	0.0247	27.480	36.55
	8	0.0003	0.0162	0.0343	0.0246	30.140	36.02
s9234-1	2	0.0006	0.0330	0.1217	0.0985	21.13	148.30
	4	0.0006	0.0330	0.1216	0.0984	75.77	149.15
	6	0.0007	0.0331	0.1213	0.0985	150.49	144.48
	8	0.0007	0.0332	0.1222	0.0984	175.38	145.13
s15850-1	4	0.0011	0.0806	0.0648	0.0650	449.35	700.43
	6	0.0011	0.0799	0.0654	0.0634	579.30	700.75
	8	0.0011	0.0798	0.0651	0.0636	682.48	705.53

cellation in a Time Warp parallel simulator, in Proc. of the 28th Annual Simulation Symposium, IEEE Computer Society Press, Apr. 1995, pp. 22–30.

[30] U. K. V. RAJASEKARAN, M. CHETLUR, G. D. SHARMA, R. RADHAKRISHNAN, AND P. A. WILSEY, *Addressing communication latency issues on clusters for fine grained asynchronous applications — a case study*, in International Workshop on Personal Computer Based Network of Workstations, PC-NOW'99, Apr. 1999.

[31] S. P. SMITH, B. UNDERWOOD, AND M. R. MERCER, *An analysis of several approaches to circuit partitioning for paral lel logic simulation.*, in In Proceedings of the 1987 International Conference on Computer Design., IEEE, NewYork, 1987, pp. 664–667.

[32] L. SOULÉ AND A. GUPTA, *Parallel Distributed-Time Logic Simulation*, IEEE Design and Test of Computers, 6 (1989), pp. 32–48.

[33] L. P. SOULÉ, *Parallel logic simulation: An evaluation of centralized-time and distributed-time algorithms*, Tech. Report CSL–TR–92–527, Computer Systems Laboratory, Stanford University, Stanford, CA, June 1992.

[34] SPECIAL SESSION:, *Recent algorithms for gate-level atpg with fault simulation and their performance assessment*, in International Synposium on Circuits and Systems, June 1985, pp. 663–698.

[35] C. SPORRER AND H. BAUER, *Corolla partitioning for distributed logic simulation of VLSI-circuits*, in Proceedings of the 7th Workshop on Parallel and Distributed Simulation, May 1993, pp. 85–92.

[36] K. SUBRAMANI, D. E. MARTIN, AND P. A. WILSEY, *SAVANT/TyVIS/WARPED: Components for the analysis and simulation of vhdl*, VHDL User's Group, (1998), pp. 195–201.

[37] J. C. WILLIS, P. A. WILSEY, G. D. PETERSON, J. HINES, A. ZAMFRIESCU, D. E. MARTIN, AND R. N. NEWSHUTZ, *Advanced intermediate representation with extensibility (AIRE)*, in VHDL Users' Group Fall 1996 Conference, Oct. 1996, pp. 33–40.

SELF-ORGANIZED CRITICALITY IN OPTIMISTIC SIMULATION OF CORRELATED SYSTEMS*

B.J. OVEREINDER[†], A. SCHONEVELD[†], AND P.M.A. SLOOT[†]

Abstract. A variety of slowly driven diffusive systems have been shown to self organize into a critical state. In this paper we study the dynamic runtime behavior of the optimistic parallel Time Warp simulation method. The method is based on an asynchronous execution of timed events. The basic problem is the out of order execution of events. A causality error results in so-called rollback of processed events until the causality error is resolved. By using the Ising spin model we show experimentally that the distribution of number of rolled back events behaves as a power-law distribution over a large range of sub-critical Ising temperatures and decays exponentially above for super-critical Ising temperatures. For critical Ising temperatures, the computational complexity of Time Warp and physical complexity of the Ising spin model are entangled and contribute both to the runtime behavior in a non-linear way.

Key words. parallel discrete event simulation, self-organized criticality, cellular automata, phase transition

AMS subject classifications. 65Cxx, 68Q80, 68U20, 82B27, 82C27, 82C44

1. Introduction. There is an increasing interest in the application of discrete event simulation to solve problems from natural sciences. In particular, problems with heterogeneous spatial and temporal behavior are, in general, most exactly mapped to asynchronous models [2, 11]. The interest in discrete event simulation is motivated by the ability of this protocol to capture the asynchronous behavior that is a qualifying characteristic of these models. Besides the aspect of asynchronicity, a general tendency is the construction of more realistic models resulting in more complex and larger simulations, which requires vast amounts of execution time. One fundamental method to reduce the execution time of large discrete event simulations is the exploitation of parallelism inherent to this class of simulations [9, 13]. Most research in Parallel Discrete Event Simulation (PDES) is focussed on protocol design; and although there are encouraging advances, none of the protocols devised thus far have been shown to perform efficiently for different applications.

We are specifically interested in the use of PDES methods, in particular the Time Warp method [6], to study dynamic complex systems modeled by Asynchronous Cellular Automata (ACA) [12, 15]. Asynchronous cellular automata differ from regular synchronous Cellular Automata (CA) in their update scheme. The danger of simply imposing an update scheme has been considered in an article by Huberman and Glance [5]. They used a spatial version of the prisoner's dilemma to show the remarkable differences between asynchronous

*This research is supported in part by the Netherlands Organization for Scientific Research, Massively Parallel Computing Priority Program under grant 95MPR01/1.

[†]University of Amsterdam, Faculty of Science, Section Computational Science, Kruislaan 403, 1098 SJ Amsterdam, The Netherlands.

and synchronous updating. In synchronous computations nothing happens for times shorter than integral values of unit time. The resulting global dynamics is mathematically described by a finite difference equation. However, in natural systems, a global clock seldom exists. In for example biological organisms and most physical systems, organisms or particles act different at different moments in time. The resulting global dynamics is usually expressed in the form of differential equations, whose solutions are not always the same as those of their finite difference counterparts.

The dynamic behavior of many natural systems is strongly imposed by the statistical (spatial) correlation between the individual components of the system. In particular, critical phenomena in Ising spin systems (i.e., magnetization in ferro-metals) or individual-based population dynamics are explained by spatial correlations [17]. The fundamental question addressed in this paper is: to which extent do the correlations in the simulation model influence the run-time behavior, or dynamics, of the PDES method. To study this phenomenon, both correlated and uncorrelated system dynamics are simulated, and their effect on causality errors in the PDES method is examined.

The complex dynamics of a particular PDES method, Time Warp, will be investigated. A highly speculative conjecture is made, that the Time Warp dynamics can be characterized as a so-called self-organized critical system [1]. Self-organized critical behavior is found in many complex systems, for example in sand-pile and earthquake dynamics. If this conjecture is true, yet another hint is given that problems in the field of parallel computing show behavior comparable to other complex systems [14].

2. The APSIS Optimistic Simulation Environment. The Amsterdam Parallel Simulation System (APSIS) is a research vehicle for both optimistic simulation protocol design and evaluation, and parallel simulation development of dynamic complex systems. Specifically, requirements for computational science applications are taken into consideration to assess the potential of PDES methods to solve problems originating from, e.g., physics, chemistry, or biology. These requirements put special constraints on the design of the simulation environment and necessitate extensions to the basic Time Warp method.

In optimistic simulation, the parallel simulation processes execute events and proceed in local simulated time as long as they have any input at all. A consequence of the optimistic execution of events is that the local clock or Local Virtual Time (LVT) of a process may get ahead of its neighbors' LVTs, and it may receive an event message from a neighbor with a timestamp smaller than its LVT, that is, in the past of the simulation process. If we allow causality errors happen, we must provide a mechanism to recover from these errors in order to guarantee a causally correct parallel simulation. Recovery is accomplished by undoing the effects of all events that have been processed prematurely by the process receiving the straggler. The net effect of the recovery procedure is that the simulation process *rolls back* in simulated time.

The premature execution of an event results in two things that have to be rolled back: (i) the state of the simulation process and (ii) the event messages sent to other processes. The rollback of the state is accomplished by periodically saving the process state and restoring an old state vector on rollback. Recovering from premature sent messages is accomplished by sending an *anti-message* that annihilates the original when it reaches its destination. A direct consequence of the rollback mechanism is that more anti-messages may be sent to other processes recursively, and allows all effects of erroneous computation to be eventually canceled.

The design and implementation of *incremental state saving* in the APSIS environment is an essential feature to effectively support the simulation of dynamic complex systems such as ACA [12]. The state vector of a spatially decomposed ACA can be arbitrarily large, that is, all the cells in the sub-lattice are part of the state vector. Incremental state saving stores not the full state vector, but saves only the changes to the state vector due to the execution of an event, which is only a small fraction of the full state. Besides efficient memory management, incremental state saving also reduces the time overhead related to memory copying.

A second necessary extension to the basic Time Warp method is optimism control, or *optimism throttling*. Uncontrolled optimism may lead to thrashing behavior, where the system experiences excessive long and/or frequent rollbacks. This results in an inefficient execution where correcting causality errors consumes more computation time than the forward simulation. An effective method to control optimism is to impose a time window on the events: the events outside the time window are delayed until the time window is updated [16]. The time window bounds the difference between the LVTs of the simulation processes and hence limits the lengths of rollback chains.

3. The Ising Spin Model. We use the Ising spin model to study the influence of spatio-temporal correlations to the dynamic behavior of the Time Warp optimistic simulation protocol. The Ising spin model is a fairly simple model that exhibits complex behavior under the control of one well-defined parameter, the temperature of the system; and as such, the model is ideal for our experiments.

The Ising spin model is a popular model of a system of interacting variables in statistical physics, and was introduced as a model for ferromagnetism. The model is defined on a grid of interacting "spins" that can either have a state "up" ($s_i = 1$) or a state "down" ($s_i = -1$). Spins on adjacent, nearest-neighbor sites, interact with each other in a pair-wise manner with a strength J. When J is positive, the energy is lower when spins are in the same direction and when J is negative, the energy is lower when spins are in opposite directions. The *magnetization* M of the system is the sum of the spin values s_i. The energy of the system is given by the following Hamiltonian:

$$E = -J \sum_{\langle ij \rangle} s_i s_j - \mu_0 H \sum_i s_i \, .$$

The first sum is over all pairs of spins that are nearest neighbors. The second term is the energy of interaction of the magnetic moments, μ_0, associated with the spins with an external magnetic field, H.

At random times, a spin is granted a chance to change the state, a so-called spin flip. The attempted spin flip, or trial, is accepted proportionally to the energy difference ΔE between the new configuration and the old configuration. Using a constant temperature (T) condition the so-called *Metropolis algorithm* can be used to simulate the dynamics. In the Metropolis algorithm, a spin flip is accepted according to the Boltzmann probability distribution $\exp(-\Delta E/kT)$. If the temperature is below a well-defined temperature, known as the critical or *Curie* temperature T_c, a spontaneous non-zero magnetization occurs ($T_c \approx 2.27$, in J/k degrees Kelvin). For temperatures $T > T_c$ the magnetization vanishes abruptly. Hence T_c separates the disordered phase for $T > T_c$ from the ferromagnetic phase for $T < T_c$.

The transition from the ordered ferromagnetic phase to the disordered paramagnetic phase is remarkably sharp, since the order parameter M approaches zero at T_c with infinite slope. Such singular behavior is an example of a "critical phenomenon." The sudden alignment of the spins as $T \to T_c^+$ (T_c^+ denotes from high T) results from correlations that are propagated extensively throughout the entire system: everything depends on everything else.

In this paper we use a discrete event simulation for the Metropolis algorithm to simulate the Ising spin model at different temperatures T. The parallel discrete event simulation in APSIS exploits the implicit spatial locality present in the Ising spin model, without altering the update history [10, 8].

4. Self-Organized Criticality. Most of the time, equilibrium systems with short-range interactions, exhibit exponentially decaying correlations. Infinite correlations, i.e., scale invariance, can be achieved by fine-tuning some parameters (e.g., temperature) to a critical value. An example of such a system is the Ising spin model of magnetization.

Besides systems exhibiting critical behavior, a large class of non-equilibrium locally interacting, nonlinear systems spontaneously develop scale invariance. Such composite systems with many interacting degrees of freedom may evolve to a critical state in which minor events may trigger a chain reaction that can affect an arbitrary large number of constituents of the system. This state is called *self-organized criticality* (SOC) [1]. The probability of spontaneously generated structures or events, further called avalanches, of many different sizes s show a power-law distribution

$$P(s) \sim s^{-\tau} ,$$

where τ is a critical exponent and most other observables of the system have no intrinsic time or length scale. This implies that we expect to observe scale invariance, or power-law scaling, in the system. The absence of intrinsic length scale is attributed to SOC, where avalanches of all sizes contribute to keep the system perpetually in a critical state. This critical state is robust with respect to any small change in the rules of the system. The size of an avalanche can be defined in different ways. It can be measured by the number of relaxation steps needed for the chain reaction to stop or the total number of sites involved in the avalanche.

Many naturally occurring systems exhibit this kind of scaling- or self-similar behavior, examples are earthquakes, stock markets, and ecosystems. The concept of SOC is developed to explain the behavior of such systems. Self-organized critical behavior was first investigated for sandpile models [1]. In this cellular automata model, a particle is dropped onto a randomly selected lattice point. When a lattice point accumulates four particles, they are redistributed to the four adjacent lattice points, or in case of edge lattice points they are lost from the grid. Redistributions can lead to further instabilities and avalanches of particles in which many particles may be lost from the edges of the lattice. The average number of particles per lattice point is the density that fluctuates about a quasi-equilibrium value. One measure of the avalanche size is given by the number of particles lost by the lattice during a sequence of redistributions, or alternatively can be given by the number of lattice points that participate in the redistribution.

5. Self-Organized Criticality in Time Warp Dynamics. As the Ising spin model is a critical system, that is for a certain parameter regime the system exhibits fractal structures, it appears an ideal simulation model to study the influence of spatial correlation on the dynamical behavior of the Time Warp protocol. The many degrees of freedom that are interacting show emergent behavior, i.e., magnetization, as the temperature of the Ising spin system approaches T_c^+. This emergent behavior is the result of the correlation length in the system that diverges as the temperature approach T_c, and is even infinite at T_c.

Self-organized critical behavior is found in *slowly driven, interaction-dominated threshold* systems (SDIDT); if a SDIDT system exhibits power-laws without any apparent tuning then it is said to exhibit self-organized criticality [7]. Interesting behavior arises because many degrees of freedom are interacting. In addition, the dynamics of the system must be dominated by the mutual interaction between these degrees of freedom, rather than by the intrinsic dynamics of the individual degrees of freedom.

An important characteristic of systems exhibiting SOC behavior is a separation of time scales. As stated before, it is required that such systems are slowly driven, that is perturbations occur on a much larger time scale than the diffusion or relaxation dynamics. The critical state in SOC systems is furthermore characterized by a stationary state where the driving forces balance the cascades. For example in the sandpile model, adding sand causes the pile to grow on the one hand, but causes avalanches on the other hand. The dynamically stationary state is obtained at the "critical point" where these two effects balance exactly.

A highly speculative analogy could be made with Time Warp: adding events causes the event rate to grow on the one hand, but causes rollbacks to occur on the other hand. Our experiments show that in Time Warp, the event rate eventually reaches a kind of stationary state with superimposed rollback cascade effects (see Fig. 5.1). We define two time scales in Time Warp: *simulation time* and *protocol time*. The simulation time in Time Warp is updated by the rate at which the system is driven. This driving rate is determined by the dynamics of the simulation. The protocol time, needed to process a rollback, is determined by machine specific parameters. A difference with conventional SOC systems and Time Warp is that there is no explicit separation of time scales, asynchronous updates and rollbacks may intervene. Also rollbacks take place on separate processors instead of an entire lattice of sites.

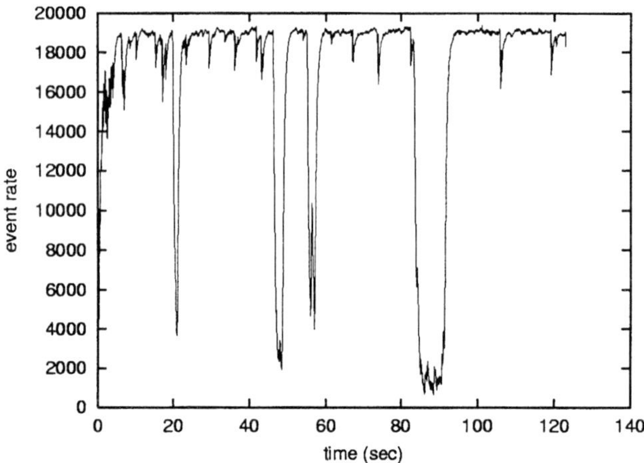

FIG. 5.1. *Event rate of one of the logical processes in a PDES Ising simulation, with number of processors* $P = 6$ *and Ising temperature* $T = 1.0$.

To understand when "relaxation" in Time Warp occurs, we first have to explain how events are processed and can give rise to rollbacks in the Ising spin model. Events in the Ising spin model are *attempts* to flip a spin in the lattice. An attempted spin flip is accepted according to the Boltzmann probability distribution. Hence, not every event results in a spin flip, or equivalently in a state change. The parallel Ising spin simulation exploits the inherent parallelism by spatial domain decomposition. Each logical process in the parallel simulation represents a subdomain of the spin lattice. Events on the subdomain boundaries that result in a spin flip are communicated with the neighboring subdomain, i.e., logical process, by way of an event message.

The so-called relaxation in Time Warp occurs whenever the following three conditions are satisfied (threshold):

- if *accepted event* and;
- *boundary event* and;
- $\exists\, i \in \text{neighborhood}(\mathit{local}) : LVT_{local} < LVT_i$;

where LVT_{local} is the simulation time on the local processor and LVT_i is the simulation time on a neighboring processor i. For Ising spin simulations, simulation time and protocol time separate at low temperatures, when there are not that many spin flips, i.e., when the acceptance ratio of the Metropolis algorithm is low. For high temperatures many spin flips are accepted, and updates and rollbacks occur at comparable time scales.

It is very important to note that, in fact, we are confronted with two kinds of critical behavior. The critical behavior of the first kind is a result of the Ising spin phase transition at the critical temperature T_c. At the Ising spin phase transition, long range spin correlations occur, that probably influence the Time Warp dynamics. We call this critical behavior of the first kind the *physical critical behavior*. The critical behavior of the second kind is conjectured, and is caused by SOC. We assume that in the low temperature Ising regime the Time Warp dynamics reaches a self-organized critical state, which we call *computational critical behavior*.

The average rollback length and rollback length distribution is studied at different temperatures in order to determine the influence of the Ising spin phase transition on the Time Warp protocol. It is expected that around the Ising spin phase transition, the long range spin correlations increase the average rollback sizes. In this way, the correlation in the Ising spin system dynamics influences the rollback behavior of the Time Warp simulation protocol. It is well known that these long range correlations result in moving islands of actively flipping spins, located in a sea of inactive spins. This separation of activity can trigger very large rollbacks whenever an active island moves over a processor boundary.

We are interested in rollback length distributions in order to do a first order check of SOC in Time Warp dynamics. Remember that for SOC systems it is well known that many observables scale as power-laws.

5.1. A First Indication of Self-Organized Criticality in Time Warp. A series of experiments are executed using the APSIS parallel simulation environment on the the Distributed ASCI Supercomputer (DAS). (Note that ASCI stands for Advanced School for Computing and Imaging—a Dutch research school.) The DAS consists of four wide-area distributed clusters of total 200 Pentium Pro nodes. ATM is used to realize the wide-area interconnection between the clusters, while the Pentium Pro nodes within a cluster are connected with Myrinet system area network technology. The experiments are performed within one single cluster, thus all communication is via the 1.28 Gbit/s Myrinet. The efficient communication primitives and thread support in the communication fabric allows for low latency, high throughput communication performance over the Myrinet network. The measured null message latency is $\pm 17\,\mu$sec, and the throughput ± 60 MB/s.

We experiment with two different grid decompositions for the parallel simulation of the Ising dynamics on a $L \times L$ square lattice: a one-dimensional "slice" decomposition and a two-dimensional "box" decomposition. Both decompositions are composed to assure optimal load balance. The individual Ising spins in the sub-lattices are aggregated into one single Time Warp logical process, resulting in a mapping of one Time Warp logical process onto a processor.

For all parameter instances of the simulation experiments we measure the average rollback length and rollback length distributions.

For the first series of experiments in this section we have fixed the lattice size to $L = 220$, the number of processors to $P = 12$, and the virtual time window (VTW) to 3000, using a "sliced" 1D decomposition. In Fig. 5.2 the average rollback lengths are shown for the temperature range [0.1–2.7]. Ideally, rollback avalanches are measured instantaneously over the system, that is over all processors used by the parallel simulation. However, the instantaneous measurement requires freezing the forward simulation, and allowing only rollbacks to occur, i.e., the relaxation in Time Warp. This would fundamentally alter the dynamics of the Time Warp protocol, and is therefore unacceptable intrusive. Hence, each processor records the local rollbacks for analysis. The rollback lengths are averaged over time for all processors. The results of three different runs are depicted in the figure. Close to the Ising phase transition ($T_c \approx 2.27$ for infinite lattices), the expected peak in the average rollback length can be observed. From this figure, three different regimes can be identified: the *physical sub-critical phase* ($< T_c$), the *physical critical phase* ($\approx T_c$) and the *physical super-critical phase* ($> T_c$).

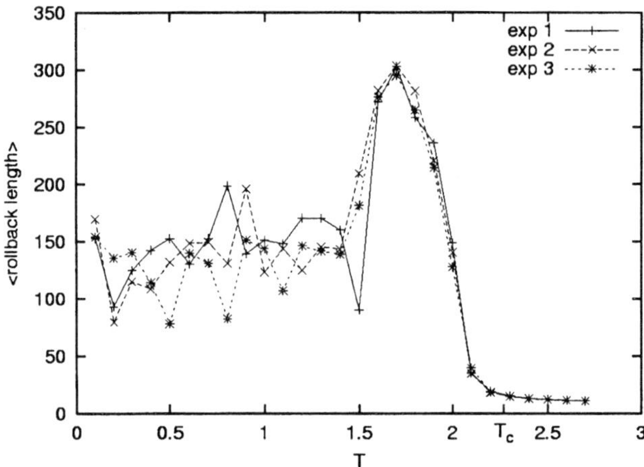

FIG. 5.2. *Average rollback length for different temperatures. For each temperature, the results of 3 experiments are shown. Using the simulation parameters* $L = 220$, $P = 12$, *and* $VTW = 3000$, *and a 1D decomposition.*

The different phases influence the rollback length distributions. In the physical sub-critical temperature regime [0.1–1.4], power-law scaling is found (see Fig. 5.3), i.e., the Time Warp dynamics appear to be in a computational critical regime. As the temperature approaches the physical super-critical regime a transition to exponential scaling can be observed (see Fig. 5.4). Close to the critical temperature, length distributions with "fat tails" (power-law distributions with exponential cutoff) develop due to the emergence of long-range spin correlations.

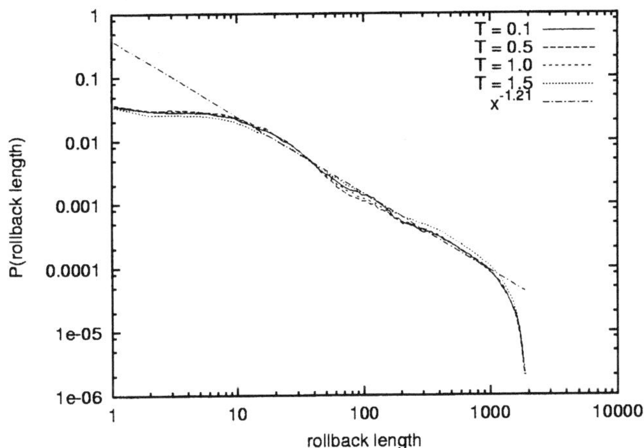

FIG. 5.3. *Rollback distribution for temperatures in the range 0.1–1.5, fitted exponent has value* -1.21 *(*± 0.01*). Using the parameters* $L = 220$, $P = 12$, *and* $VTW = 3000$, *and a 1D decomposition.*

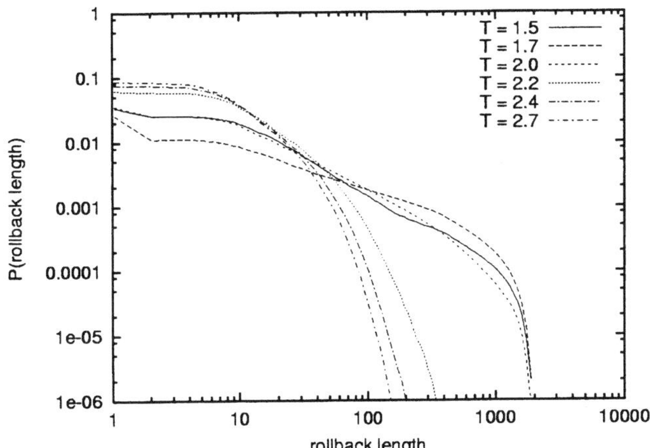

FIG. 5.4. *Rollback distribution for temperatures in the range 1.5–2.7. Using the parameters* $L = 220$, $P = 12$, *and* $VTW = 3000$, *and a 1D decomposition.*

The scaling exponent α in the physical sub-critical regime seems to be universal for all temperatures in this regime. From the experimental data a power law with exponent $\alpha = -1.21$ is fitted. Because the rollback length distribution obeys power-law scaling, we conjecture that the rollback dynamics are in a SOC regime.

The slope of the power function flattens out in the range of its lower magnitude rollbacks with length $1 - 10$. This flattening, rather than remaining double log-linear, occurs because discreteness effects of the rollback lengths come into play. As the rollback cascades approach the size of what is assumed to be the size of a real-world system's component particle, in our study an event, it becomes impossible for the fractal pattern to repeat at this scale [1, 3, 4]. The exponential cutoff is further explained in Section 5.3.

For the 2D decomposition we repeat the same set of experiments as for the 1D case, again with $L = 220$ and $P = 12$. As for the 1D case, we observe a peak in the average rollback lengths around T_c (see Fig. 5.5). Again a transition from power-law scaling to exponential scaling is observed (see Figs. 5.6 and 5.7). In the physical sub-critical regime we find $\alpha = -1.25$, slightly larger as in the 1D case. This could be the result of slightly shorter distances between processor partitions, enabling a faster propagation of cascading rollbacks.

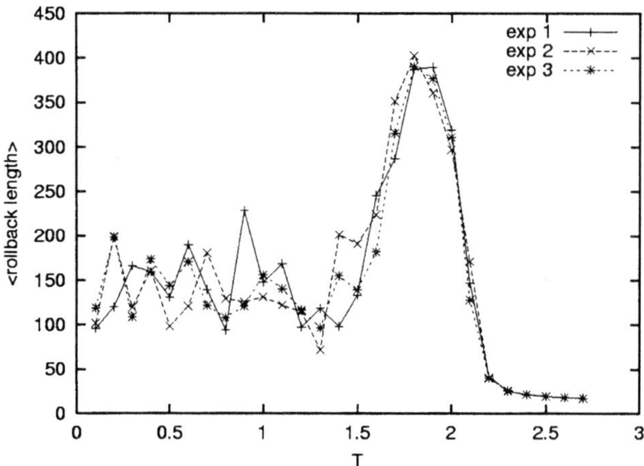

FIG. 5.5. *Average rollback length for different temperatures. For each temperature, the results of 3 experiments are shown. Using the parameters $L = 220$, $P = 12$, and $VTW = 3000$, and a 2D decomposition.*

In the next series of experiments, different parameters are varied. Note that two different processes intervene: the Ising simulation process and the Time Warp process. An important parameter for both processes is the lattice size. Due to finite size effects, increasing the lattice sizes causes the Ising spin phase transition point T_c to shift. For the Time Warp process, the probability to select a boundary cell decreases for increasing lattice sizes. If the number of processors is increased and the lattice size is kept fixed, the probability to select a boundary cell increases. Therefore we experiment with different numbers of processors. The virtual time window is a very important parameter that determines the maximum size of the rollbacks. This parameter is studied in the last series of experiments.

5.2. Varying the Number of Processors. To study the influence of the number of processors on the rollback length distribution in the SOC or computational critical regime, a series of experiments with $P = \{4, 8, 12, 24\}$ using

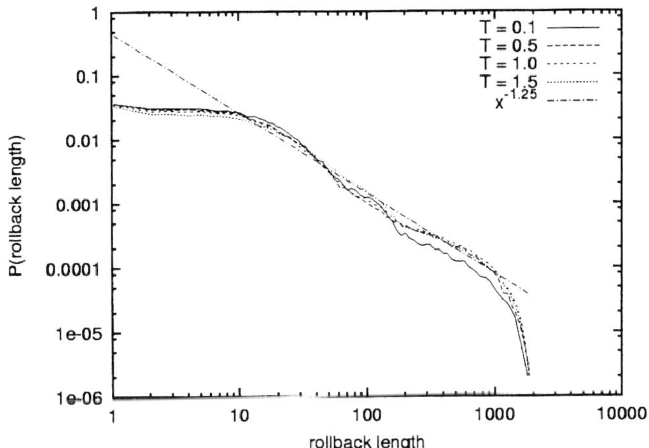

FIG. 5.6. *Rollback distribution for temperatures in the range 0.1–1.5, fitted exponent has value* -1.25 *(*± 0.02*). Using the parameters* $L = 220$*,* $P = 12$*, and* $VTW = 3000$*, and a 2D decomposition.*

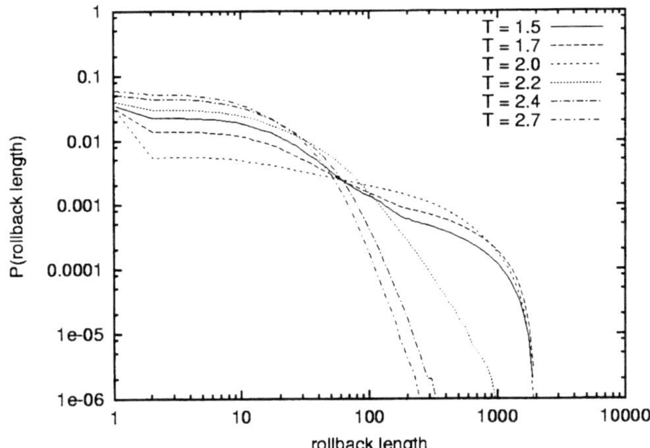

FIG. 5.7. *Rollback distribution for temperatures in the range 1.5–2.7. Using the parameters* $L = 220$*,* $P = 12$*, and* $VTW = 3000$*, and a 2D decomposition.*

a 2D decomposition has been performed. The lattice size has been fixed to $L = 220$. The rollback distributions of Ising spin simulations at $T = 1.0$ are shown in Fig. 5.8. Similar results are seen for other temperatures T in the SOC regime. The results indicate that for increasing P the rollback length distributions converge. Again, the scaling exponent α is not influenced by increasing P.

The average rollback lengths for different processors in the range $T = [0.1 - 2.7]$ are shown in Fig. 5.10. For 8, 12 and 24 processors, again, a peak around

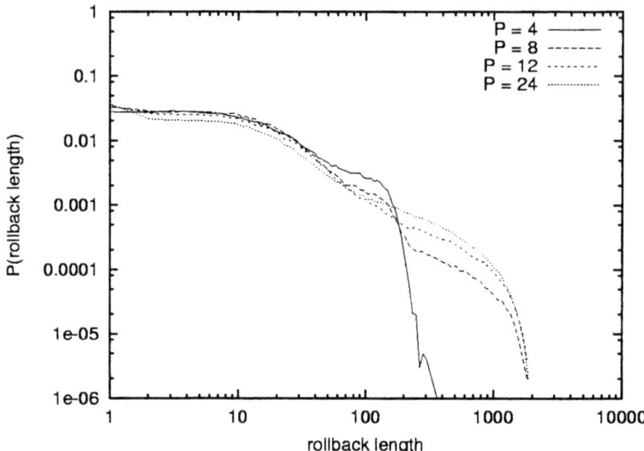

FIG. 5.8. *Rollback distributions for $P = \{4, 8, 12, 24\}$ at $T = 1.0$ using the parameters $L = 220$, $VTW = 3000$, and a 2D decomposition.*

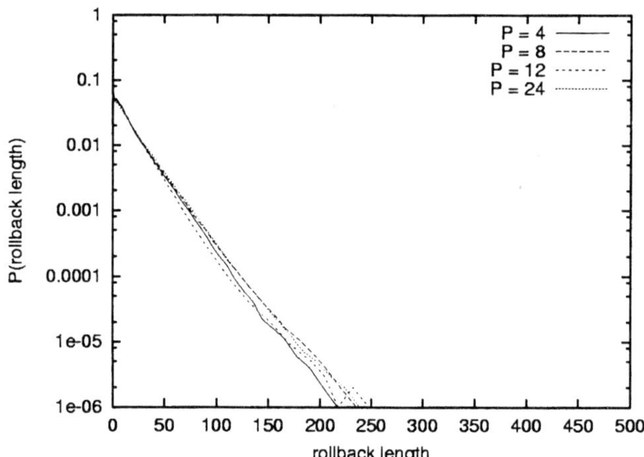

FIG. 5.9. *Rollback distributions for $P = \{4, 8, 12, 24\}$ at $T = 2.7$ using the parameters $L = 220$, $VTW = 3000$, and a 2D decomposition. Note that the horizontal axis has a linear scale.*

T_c can be distinguished. For $P = 4$, this peak is not present. Apparently, the critical Ising spin dynamics do not reduce the performance of the Time Warp protocol if only 4 processors are used.

From Fig. 5.8 we observe that for $P = 4$ the rollback length distribution does not show power-law scaling behavior. The cutoff size of the rollback length distributions for processor numbers $P = 8$ and larger is determined by the virtual time window (VTW) size. (This effect is in more detail presented in Section 5.3.) In the low temperature regime the *average* rollback length increases with the number of processors P (see Fig. 5.10).

This is not valid anymore in the high T regime, where the rollback length distributions approximately collapse (see Fig. 5.9) to the same exponentially decreasing distribution. As a direct consequence, the average rollback lengths will collapse (see Fig. 5.10). In the high T regime the rollback lengths are not influenced by P as in the low T regime. The processors are synchronized frequently in this regime, due to a high acceptance ratio of flipped spins. Therefore, there is hardly any real time to build large simulation time differences between the processors, resulting in only small rollback lengths.

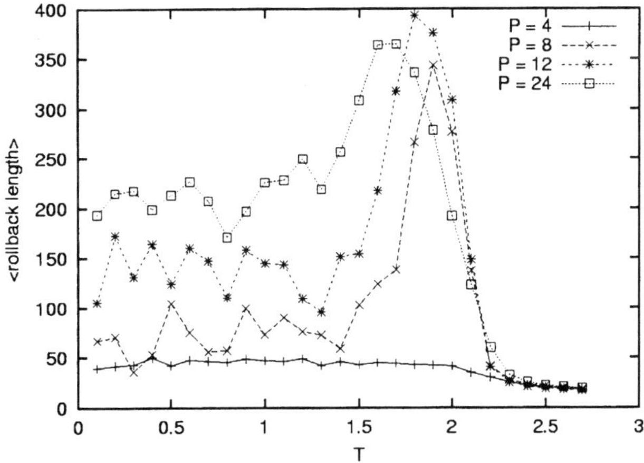

FIG. 5.10. *Average rollback lengths for* $P = \{4, 8, 12, 24\}$ *for varying* T *using the parameters* $L = 220$, $VTW = 3000$, *and a 2D decomposition.*

It is interesting to compare the runtimes for different P in the same temperature range. The results are presented in Fig. 5.11. From the figure one can derive that in the low T regime, the runtime scales down if more processors are used. This is also valid for the high temperature regime. Around T_c, we find non-trivial scaling for the different processors. Obviously, using only 4 processors gives the best result, which could be expected from the significantly lower average rollback length in this regime. Using 12 processors gives the worst results in this case.

For the high and low temperature regimes $P = 24$ gives the best performance results. Even though, in the low T regime, the average rollback length is maximal for $P = 24$, the extra overhead is beneficially applied to efficiently exploit the parallelism present in the simulation.

Although the average rollback lengths in the low T regime are much larger than the average rollback lengths in the high T regime, the execution times are comparable. This is a result of the frequency of rollback events. In the low T regime this frequency is much lower, due to the reduced acceptance probability of spin flips. It seems that, the average rollback lengths and the rollback frequency

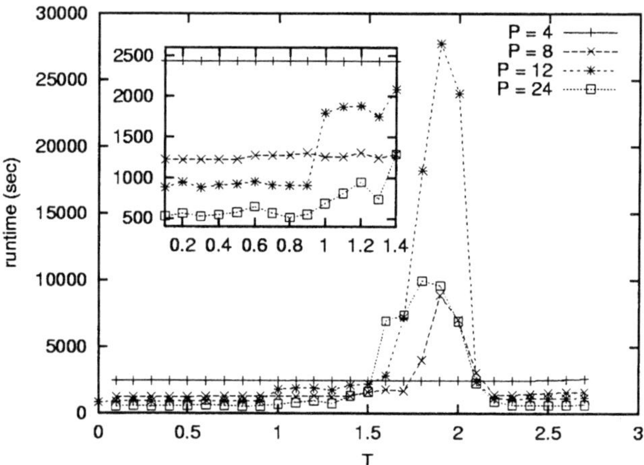

FIG. 5.11. *Run times for $P = \{4, 8, 12, 24\}$ for varying T using the parameters $L = 220$, $VTW = 3000$, and a 2D decomposition.*

are balanced to approximately similar execution times for the low and high T regimes.

5.3. Different Virtual Time Window Sizes. An important parameter of the Time Warp protocol is the so-called *virtual time window* (VTW). This parameter controls the asynchronicity of the simulation. It specifies the maximum difference between the local virtual time and the global virtual time (the minimum of all local virtual times). It is expected that this parameter greatly influences the rollback dynamics. For the experiments presented in this section we have varied the VTW parameter while keeping all other parameters fixed ($P = 12$ and $L = 220$).

In Fig. 5.12 the rollback distributions are depicted for $T = 1.0$ for experiments with VTW parameters in the range $[750, 6000]$. Obviously, a small virtual time window decreases the maximum rollback length.

In Fig. 5.13 the rollback distributions for VTWs around 3000 are shown. There is a transition from $VTW = 2750$ to $VTW = 3000$. The VTW values $\{3000, 3250, 3500\}$ produce similar rollback length distributions, while $VTW = 4000$ deviates.

From Fig. 5.14 it can be observed that the peak of the average rollback length shifts and broadens for increasing VTW. This effect is caused by the Ising dynamics. Large virtual time windows effectively result in a more pronounced influence of the finite sub-lattices (decomposed over the 12 processors). Due to the increased asynchronicity for larger time windows the sub-lattices are effectively loosely coupled and act more like individual Ising spin lattices. It is a well known fact in Ising spin simulations that decreasing the lattice size results in a broadening and shifting of the spin correlation peak around T_c.

The average rollback lengths roughly decreases for decreasing virtual time

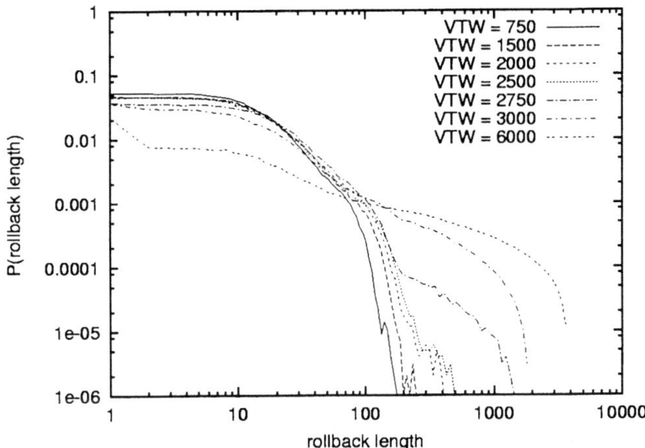

FIG. 5.12. *Rollback distributions for VTW* $= \{750, 1500, 2000, 2250, 2500, 2750, 3000, 6000\}$ *at* $T = 1.0$ *using the parameters* $L = 220$, $P = 12$, *and a 1D decomposition.*

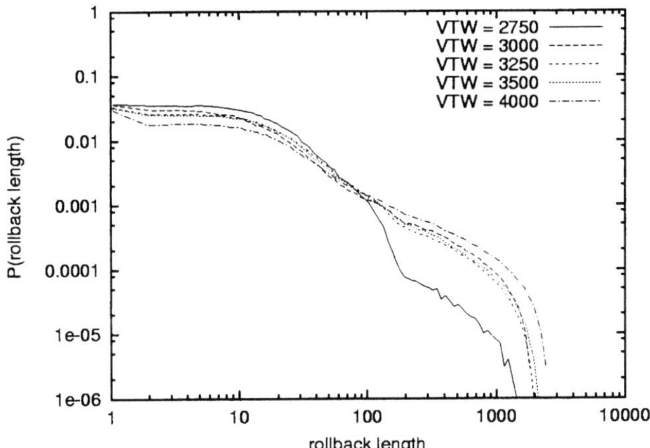

FIG. 5.13. *Rollback distributions for VTW* $= \{2750, 3000, 3250, 3500, 4000\}$ *at* $T = 1.0$ *using the parameters* $L = 220$, $P = 12$, *and a 1D decomposition.*

window size (see Fig. 5.14). This is a result of the Time Warp dynamics. Small virtual time windows only allow for a small build up of local virtual time differences.

For smaller VTW values the average rollback lengths are comparable over the entire temperature range. For these values it can be concluded that the Time Warp dynamics are not constrained to the details of the Ising dynamics. The increased synchronization frequency disables the build up of large time differences.

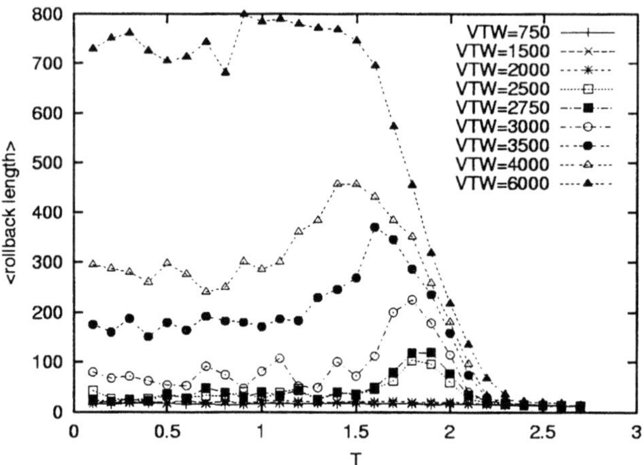

FIG. 5.14. *Average rollback length for* $VTW = \{750, 1500, 2000, 2250, 2500, 2750, 3000, 3500, 4000, 6000\}$ *for varying* T *using the parameters* $L = 220$, $P = 12$, *and a 1D decomposition.*

Somewhere there is a crossover point where increasing the maximum rollback lengths (by increasing VTW) does not improve the progress in simulation time due to the increased protocol overhead. For this specific simulation instance it seems that $VTW = 2750$ is optimal for the low and high temperature regime (see Fig. 5.15). For the regime around T_c it is almost optimal. Note that $VTW = 3000$ is comparable to $VTW = 2750$ in the low and high T regimes, while around T_c, $VTW = 3000$ produces significantly higher execution times. Hence the run times are highly susceptible for VTW around T_c, as a consequence of the critical Ising dynamics.

In contrast to the disappearance of the average rollback length peak in Fig. 5.14 for increasing VTW, a peak remains in the runtime curves. This can be explained from reduced rollback frequencies in lower temperature regimes. The increase of the runtimes around T_c with increasing VTW can be explained from the increased average rollback lengths (see Fig. 5.14) and the fat tail in the rollback size distribution around T_c for large virtual time windows (data not shown).

6. Conclusions. In this paper we have intensively studied the dynamical behavior of the Time Warp protocol for parallel discrete event simulations. As a simulation case we considered the Ising spin model, which is basically a cellular automata model. A property of the Time Warp protocol is the appearance of so-called rollbacks whenever a causality error occurs. This rollback mechanism can trigger a cascade of events that need to be undone. The local rollbacks are recorded by the simulation process, as the instantaneous global cascaded rollbacks cannot be administered without unacceptable intrusion to the Time Warp dynamics. It is known that so called slowly-driven, interaction-dominated

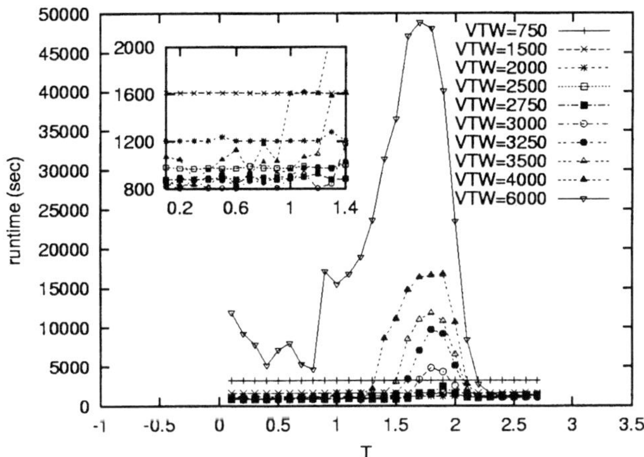

FIG. 5.15. *Runtimes for* $VTW = \{750, 1500, 2000, 2250, 2500, 2750, 3000, 3250, 3500, 4000, 6000\}$ *for varying* T *using the parameters* $L = 220$, $P = 12$, *and a 1D decomposition.*

threshold (SDIDT) systems can exhibit power laws without any apparent tuning. The specific feature of these dynamical systems is called self-organized criticality (SOC).

From the inset in Fig. 5.11 we can see that the optimistic simulation of the Ising spin system scales linearly with the number of processors for low temperatures $T = [0 - 1.5]$. However, it is found that the Ising spin phase transition influences the rollback behavior, and consequently the runtime. Around the critical temperature (physical critical behavior) T_c, the average rollback lengths increase dramatically, as well as the simulation runtimes, due to long-range spin correlations. The non-trivial scaling in runtimes around the critical temperature shows in Fig. 5.11 that the best performance is obtained with only 4 processors. For physical sub- and super-critical temperatures the simulation runtimes approximately coincide.

For the rollback dynamics three different phases can be distinguished: *physical sub-critical*, *physical critical*, and *physical super-critical* rollback length scaling behavior. In the sub-critical regime the scaling behavior appears to behave like a power-law, with exponents independent of the temperature. In this regime we conjecture that *computational critical* (SOC) behavior appears. Note that in the sub-critical regime, the Ising spin system is *not* in a critical state. Hence, the application does not need to exhibit critical behavior for the Time Warp simulation protocol to show SOC. Around the critical phase large rollback lengths become more abundant due the long-range spin correlations. Here the computational complexity and the physical complexity are entangled and contribute both to the runtime and rollback behavior in a non-linear way. In the physical super-critical phase a negative exponential distribution of the rollback lengths is observed.

Obviously a lot of work remains to be done in the study of physical- and computational critical behavior in Time Warp. Complementary studies are necessary, such as the simulation of the dynamic behavior of Time Warp and the Ising spin simulation. The simulation of the model of Time Warp and Ising spin simulation allows for a full statistical analysis of the runtime behavior. In the current study, only local rollbacks are administered. However the sizes of the cascaded rollbacks is another observable of the system and should show power-law scaling behavior.

Besides Ising spin models, other application systems with (long-range) spatio-temporal correlations should be considered for evaluation of SOC behavior. In particular, spatial decomposed problems like population dynamics, forest fire, or damage spreading models are potential candidates. The study of other simulation applications with spatio-temporal correlations will further increase our insight into the power-law rollback behavior and complexity entanglement in the Time Warp simulation protocol.

The results presented in this paper are, to our knowledge, the first series of experiments that have ever been conducted to study the influence and the appearance of critical behavior in Time Warp. The entanglement of the computational and physical complexity, and their non-trivial contribution to the runtime behavior might have consequences for other optimistic simulations.

REFERENCES

[1] P. Bak, C. Tang, and K. Wiesenfeld, *Self-organized criticality*, Physical Review A, 38 (1988), pp. 364–374.

[2] H. Bersini and V. Detours, *Asynchrony induces stability in cellular automata based models*, in Proceedings of the IVth Conference on Artificial Life, Cambridge, MA, July 1994, pp. 382–387.

[3] G. Brunk, *Understanding self-organized criticality as a statistical process*, Complexity, 5 (2000), pp. 26–33.

[4] V. Frette, K. Christensen, A. Malthe-Sørensen, J. Feder, T. Jøssang, and P. Meakin, *Avalance dynamics in a pile of rice*, Nature, 379 (1996), pp. 49–52.

[5] B. Huberman and N. Glance, *Evolutionary games and computer simulations*, Proceedings of the National Academy of Sciences of the United States of America, 90 (1993), pp. 7716–7718.

[6] D. Jefferson, *Virtual time*, ACM Transactions on Programming Languages and Systems, 7 (1985), pp. 404–425.

[7] H. Jensen, *Self-Organized Criticality*, Cambridge University Press, 1998.

[8] G. Korniss, Z. Toroczkai, M. Novotny, and P. Rikvold, *From massively parallel algorithms and fluctuating time horizons to non-equilibrium surface growth*, Physical Review Letters, 84 (2000), pp. 1351–1354.

[9] M. Livny, *A study of parallelism in distributed simulation*, in Proceedings of the 1985 SCS Multiconference on Distributed Simulation, San Diego, CA, Jan. 1985, pp. 94–98.

[10] B. Lubachevsky, *Efficient parallel simulation of dynamic Ising spin systems*, Journal of Computational Physics, 75 (1988), pp. 103–122.

[11] E. Lumer and G. Nicolis, *Synchronous versus asynchronous dynamics in spatially distributed systems*, Physica D, 71 (1994), pp. 440–452.

[12] B. Overeinder and P. Sloot, *Application of Time Warp to parallel simulations with asynchronous cellular automata*, in Proceedings of the 1993 European Simulation

Symposium, Delft, The Netherlands, Oct. 1993, pp. 397–402.

[13] ———, *Parallel performance evaluation through critical path analysis*, in High-Performance Computing and Networking (HPCN Europe '95), no. 919 in LNCS, Springer-Verlag, May 1995, pp. 634–639.

[14] A. Schoneveld, J. de Ronde, and P. Sloot, *On the complexity of task allocation*, Complexity, 3 (1997), pp. 52–60.

[15] P. Sloot and B. Overeinder, *Time-Warped automata: Parallel discrete event simulation of asynchronous CA's*, in Proceedings of the Third International Conference on Parallel Processing and Applied Mathematics, Kazimierz Dolny, Poland, Sept. 1999, pp. 43–62.

[16] L. Sokol, B. Stucky, and V. Hwang, *MTW: A control mechanism for parallel discrete simulation*, in Proceedings of the 1989 International Conference on Parallel Processing, vol. III, Minneapolis, MN, Aug. 1989, pp. 250–254.

[17] H. Stanley, *Scaling, universality, and renormalization: Three pillars of modern critical phenomena*, Reviews of Modern Physics, 71 (1999), pp. S358–S366.

SOME OWNERSHIP MANAGEMENT ISSUES IN DISTRIBUTED SIMULATION USING HLA/RTI *

FARSHAD MORADI[†‡], RASSUL AYANI[†§], AND GARY TAN [†]

Abstract. To study the High Level Architecture (HLA) and the services that are provided by the Runtime Infrastructure (RTI), in particular Object Management and Ownership Management, we have developed a distributed air traffic controller simulator. In our simulation model, each airport is represented by a federate and controls a number of aircraft. The control of the aircraft is transferred among airports as the aircraft fly to different airports. In this paper we discuss two different approaches that we have used to facilitate the exchange of ownership of aircraft attributes among federates, namely the *pull* and the *negotiated push* method. We present a comparison of the two methods and also discuss the problems associated with each method and our approach to resolving them. These problems include the *oscillation effect*, which causes aircraft attributes to be pulled back and forth between federates, and the pending attribute acquisition requests, which result in loss of aircraft (or unattended aircraft attributes). We have experienced that in such scenarios, the push method is more efficient and accurate. We also present our experiences and observations from our experimentation with the RTI. We have noticed some shortcomings in the current RTI interface specification that we will discuss in the paper.

Key words. HLA, RTI, Ownership Management, Object Management, Air Traffic Control

1. Introduction. The High Level Architecture (HLA) Runtime Infrastructure (RTI) is a distributed operating system which provides a set of commonly required services to simulation systems (federates) participating in a joint HLA simulation exercise (federation). It supports federates by providing a set of management services. These services are divided into six categories, Federation Management, Declaration Management, Object Management, Time Management, Data Distribution management and Ownership Management services, as described in [1, 2, 3]. Much research has been conducted on the various management services, notably Data Distribution Management and Time Management [4 , 5 , 6, 7].

In order to study HLA and the services provided by the RTI, and to conduct experiments, a test platform has been designed and developed. The test-platform is a non-military HLA federation, which simulates air traffic control (ATC), in a distributed environment using the DMSO/RTI 1.3 [8]. The ATC test-bed is a distributed simulation, where each airport is represented by a federate and controls a number of aircraft, which are identified as simulation objects, within its radar range. Furthermore each airport federate is responsible for updating and publishing information about the aircraft that it controls.

One of our main objectives has been to investigate issues concerning consistency of objects' data. These issues can be experienced when for instance

---

* This research is supported by NUS Academic Research Grant RP3981671

[†] School of Computing, National University of Singapore, Singapore 119260.

[‡] On leave from Swedish Defence Research Agency (FOI).

[§] Visiting Professor from KTH, Sweden.

data is transferred between models with different aggregation levels or when the ownership of attributes of objects is transferred between federates, i.e., ownership management of objects. Hence we mainly used our test-bed as a means of studying this facet of HLA, namely Ownership Management (OWM) and have performed a thorough investigation of its services.

Ownership Management (OWM) provides federates with the possibility to exchange ownership of object attributes among themselves. Each attribute characterizes some aspect of the object's state and all together they represent the total state of an object. OWM enables federates to share and transfer the ownership of attributes of object instances and the responsibility for updating and publishing of those attributes. There are however, some restrictions associated with these capabilities. For example no single attribute of an object instance can be owned by two federates at any time. Besides that federates can share the attributes of an object instance among themselves.

OWM is also one of the services provided by the ALSP Infrastructure Software (AIS) [9]. ALSP (Aggregate Level Simulation Protocol) is the DoD's standard for interfacing constructive, time managed simulations and the AIS is analogous to the RTI in HLA.

Although the full potential of these services has not been explored, there are different scenarios where migration of attributes is very useful. One of the most common usages of these services has been within the aggregation/disaggregation processes, where the ownership of object attributes is transferred between simulation models with different levels of aggregation. In these processes the same object can be controlled (simulated) by different simulation models at different time periods. Hence, the ownership of the object (or object attributes) has to be exchanged among federates.

There are also scenarios where these services have been used to cut down the network communication cost and latency by transferring the ownership of objects, which communicate frequently, to the same federate [10]. The objective of these implementations has been to minimize the cost of sending attribute updates or interactions over the network and saving time and network resources. This is particularly essential in time critical simulations, where some objects need to get quick access to information about the state of other objects.

Another reason for utilizing the OWM services would be that the nature of the underlying system requires that the ownership of some of the attributes of an object is exchanged among federates, as is the case in our ATC system.

Recently, several researchers have suggested some alternative approaches to the object ownership transfer in HLA [11, 12]. In this paper we focus mainly on ownership transfer of certain attributes of an object, i.e., attribute ownership management (AOM), rather than the ownership of the object as a whole as discussed by Li et al. [11, 12]. For instance, in an ATC, the owner of an aircraft may want to transfer the ownership of the "position" attribute to another airport, but not the attributes of the passengers of the aircraft. In this paper, both the terms "object ownership" and "attribute ownership" refer to attribute ownership.

We describe our experiences from working with OWM services, and also present and evaluate different methods for transferring attribute ownership among federates, namely *push* and *pull*. A comparison of thcsc two methods is also given.

Furthermore, we describe several approaches to handle detection and discovery of objects and how to transfer attribute ownership, in the context of the ATC environment. We present some of the object management problems that may arise when simulating scenarios such as the ATC where objects' attributes are passed continually from one federate to another, and how these problems can be resolved. We also discuss some of our experiences from working with the RTI.

2. Ownership Management. In the previous section we explained the idea behind OWM and the reasons why it has been incorporated in HLA Interface Specification and the RTI. In this section we focus on different methods for transferring attribute ownership among federates and describe how the OWM services have been implemented and used in the RTI.

Basically there are two approaches for transferring attribute ownership among federates. One is the *push* approach, where the federate that owns the attributes initiates the ownership exchange by informing the RTI of its willingness to give up the ownership. In the other method the federate that wishes to take over the ownership of a given attribute sends an acquisition request to the RTI and initiates the process. This is called the *pull* approach.

In the former method the federate that is responsible for updating and publishing the attributes of an object instance (the owner) decides to give up its rights (ownership) and initiates the procedure. The push procedure can be done in two different ways, unconditional or negotiated (conditional). In the unconditional ownership release, the owner federate informs the RTI that it wishes to give up the ownership or responsibilities for attributes without knowledge of any existing receivers. The federate can immediately release its responsibilities by invoking the "unconditionalAttributeOwnershipDivestiture" method. If there is no federate interested in those attributes they will be left unowned.

In the "negotiated push" approach (figure 2.1) however, the owner federate makes sure that there is an interested party among the other federates before it releases the ownership. In this way the federate is assured that the attributes are not left unowned. This handshaking process is started when a federate calls the RTI by invoking the "negotiatedAttributeOwnershipDivestiture" method. The RTI in turn notifies all federates that are eligible to take over the ownership of the given attributes by sending a "requestOwnershipAssumption" to the potential receivers. If there is a federate that wishes to acquire the ownership of the offered attributes, it can inform the RTI by responding through the invocation of the "attributeOwnershipAquisition" or the "attributeOwnershipAquisitionIfAvailable" methods. Upon receiving these calls the RTI sends a callback and notifies the owner (first federate) that it does not own the attributes anymore and it should stop updating them. It also sends a notification to the new owner

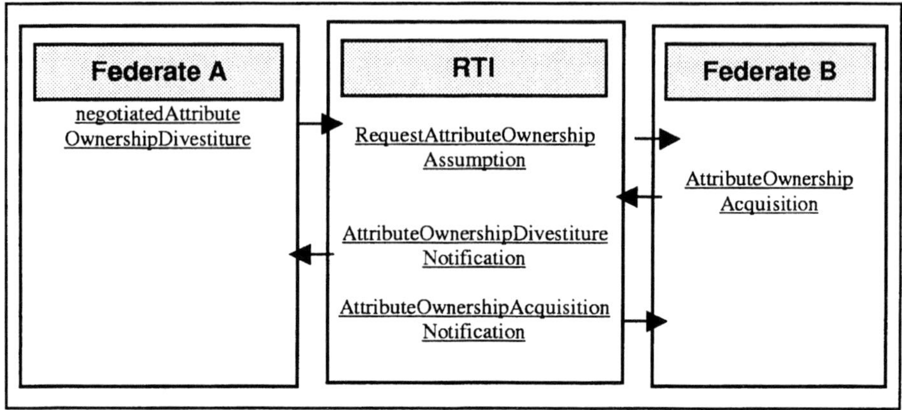

FIG. 2.1. *Procedure for transferring ownership of attributes from federate A to federate B, where federate A initiates the process but does not give away the ownership until federate B accepts to take over the attributes (negotiated push).*

so that it can start updating and publishing the attributes.

The pull method, as shown in figure 2.2, is started when a federate sends an acquisition request to the RTI. Subsequently the RTI informs the owner federate by sending a "requestAttributeOwnershipRelease" to it. The owner federate may respond to this request by invoking the "attributeOwnershipReleaseResponse" and telling the RTI that it has accepted to release the responsibility. Upon receiving this information, the RTI will notify the new owner that the cycle is completed.

In addition to these, the Ownership Management service facilitates some supporting functions, e.g., mechanisms for identifying the current owner of a particular attribute or cancelling an acquisition request or a negotiated divestiture. (For more information about Ownership Management services see HLA Interface Specification, [8]).

3. Air Traffic Control Simulation (ATC). In order to experiment with different services provided by the RTI, we needed a suitable test federation. Since HLA (for historical reasons) has been mostly used in defense applications we decided to use a non-military example. We needed a distributed simulation test-bed that would be sufficiently simple so that we could quickly pass the programming stages and get on with the experiments. At the same time it has to be complex enough to use all (or most of) the RTI services, especially Data Distribution Management [13], and Object Ownership Management [14]. It was also desirable to have federates that could simulate a large number of simulation objects which move between federates, so that federates would exchange the responsibility for updating their attributes. With all this in mind we decided to develop and implement an air traffic control model that could serve our purpose.

The air traffic federation that we developed is a distributed simulation,

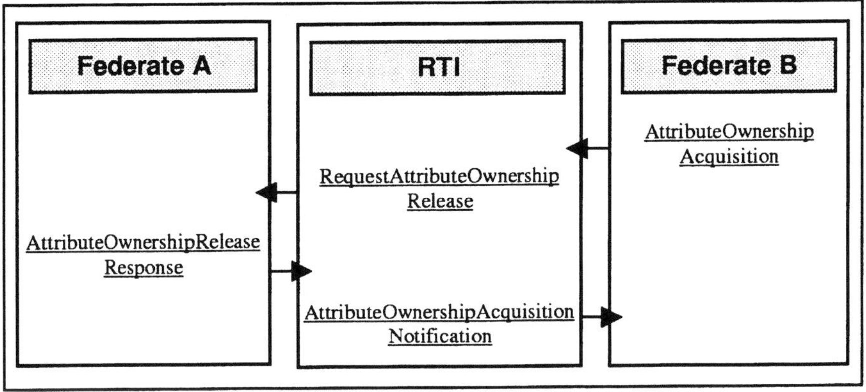

FIG. 2.2. *Procedure for transferring ownership of attributes from federate A to federate B, where federate B initiates the process (pull approach).*

where all federates are airports located at different geographical positions (figure 3.1). Each airport is initially assigned a number of aircraft, which it controls. Furthermore the airports are responsible for updating and publishing information about aircraft that they control. Aircraft fly from one airport to another, land, choose a new destination and fly again to the new destination. Whenever an aircraft approaches the boundaries of its controlling airport (or the current owner), the ownership of the aircraft is transferred to a new airport, which becomes the new controller and will be responsible for updating the attributes of the aircraft.

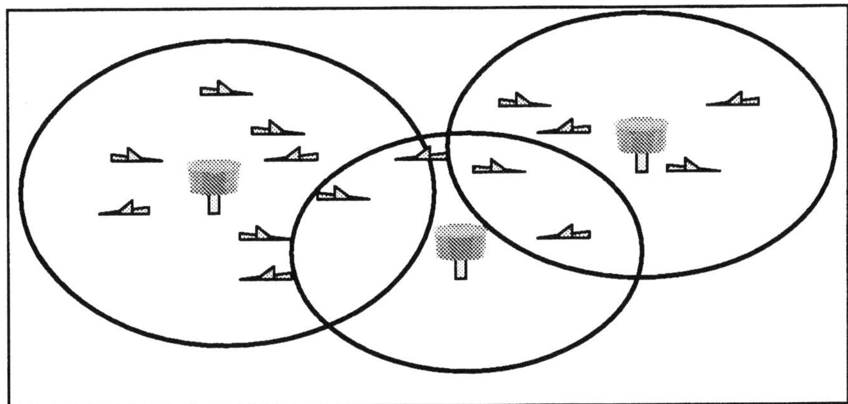

FIG. 3.1. *A simplified sketch of an Air Control scenario with airports located at different locations and aircraft flying between the airports.*

In a real air traffic situation each airport has a set of radar, which cover a portion of the air space that they monitor (territory). Hence each airport detects all traffic that passes through the region that is covered by its radar. Furthermore, in a real scenario an aircraft has to be in contact with an airport along its route from the time it takes off till the time it lands. It starts with the airport of origin and when the aircraft is to leave the covering region of that airport, another airport along the route (whose covering region overlaps with the first one) will take over the contact (responsibility or monitoring) through some handshaking procedures. This will go on until the aircraft reaches the destination and lands. That is the way the air traffic is coordinated so that no aircraft is left unmonitored at any time [1].

Consequently in our model we have a three-dimensional routing space with x, y and z coordinates. Each airport has two different radar ranges that define its publication and subscription regions in this routing space. As is in the real case the publication and subscription regions of neighboring (adjacent) airports overlap. Initially each aircraft is given a destination which could be any of the federates in the simulation.

In reality there is a certain handshaking procedure that all airports follow in order to transfer and exchange the control of airplanes among themselves, but for experimental reasons we decided not to follow those rules exactly to the point. This way we could experiment with different methods and explore the possibilities of the RTI.

Our implementation of ATC is particularly interesting for studying OWM, because unlike many other applications, where the ownership transfer seldom occur, these services are highly stressed and the ownership exchange happens very frequently in ATC.

Another simplification that we made was with the air space regions. Unlike the air space regions of the real airports, where the shape of the region is not regular and depends on the territorial boundaries of the airports or the countries, the regions in our model are regular cylindrical or cubic portions of the airspace.

4. Ownership Management in ATC. As explained in section 2, there are two main approaches to transfer of ownership, push and pull. In this section we describe our implementation of these two methods and discuss the advantages and disadvantages of each method.

4.1. Pull. In this method, as described earlier, when an airport discovers that an aircraft has entered its air space (subscription region), (if some conditions are satisfied) it will initiate the ownership exchange process by sending a request to the RTI to take over the control of the aircraft. This is naturally done while the aircraft is still inside the publishing region of the owner airport, since the subscription and publishing regions of the airports overlap each other. Subsequently, the RTI sends a callback to the owner federate and informs it that there is an interested party. When the owner federate receives the callback from the RTI, it stops updating the attributes in question and informs the RTI

[1]Part of the regulations we have used in our simulation model is extracted from the annual publication of the Civil Aviation Authority of Singapore (CAAS), given in [15].

that the attributes could be taken over. Hence the RTI gives the requesting federate the green light by sending a callback and telling it to start updating the attributes and this concludes the process.

4.1.1. Undesirable Oscillation effect. There are however some potential problems when using this approach. For example, after a successful ownership transfer, if the aircraft in question is still in the intersection region covered by the new owner and the former owner, the former owner may then discover the aircraft and issue a request to acquire the ownership again. If the new owner agrees to transfer the ownership, this will lead to unnecessary changing of owners. This situation does not arise in the real system since the former owner has sufficient information to recognize the aircraft and will not send out such a request. However, this unnecessary ownership change can occur in the simulated ATC, because federates do not maintain much history information about the objects, if any, and hence will not be able to recognize the aircraft. This problem could either be solved at the requesting federate's side or at the owner's side. This means that either the new owner should refuse to give up the ownership or the old owner should not try to take over the ownership. Experimentation has shown that working solely on either side would cause some additional problems.

One solution for solving the problem at the former owner's side is to disable it from sending a new request to take over attributes that it had just passed over to another federate. One may use time constraints so that each federate has to wait for a certain period of time before it tries to regain the previous ownership. However, it is hard to know how long this time period should be. It can depend on the size of the regions and their intersection areas, and aircraft speed.

Yet another solution could be to use some geographical calculation to see whether it is necessary for a federate to acquire the responsibility for updating an aircraft or not. This solution is very difficult to implement and may not even work. One may claim that if the aircraft is moving away from a federate, then it is not reasonable that this federate should take over the ownership. However, it is easy to construct examples where such a request is reasonable.

Solving the problem at the owner's side means that when the federate receives a "requestAttributeOwnershipRelease" message from the RTI, it does not reply to it and simply ignores it (since there is no means in the RTI to reject an attribute ownership release request). However, this will cause another problem, because we now have a pending request that has been left unanswered. So, the next time the aircraft turns back or just somehow enters the former owner's territory, the federate will send a new request but the RTI will not forward this to the owner, because there is already an outstanding request there to which it has not replied. Consequently this may cause a situation where some object attributes are left unattended (or loss of aircraft).

This situation could be avoided if the owner federate could somehow tell the requesting federate that its request has been rejected. In this way the requesting federate would cancel its request and ask the RTI to discard it. Unfortunately there is no mechanism in the RTI to inform a requesting federate that its request has been rejected. In other words the three-way handshaking mechanism in the OWM services of the RTI is not complete and sufficient. We will present our solution for completing this process in section 6.3. The problem with the pending request and our approach for solving it will be discussed in the next sub-section.

We have devised a solution to overcome the oscillation effect, which will work for both sides. The solution we implemented was to differentiate between the subscription and the publication regions of the airports and give them two different sizes. The regions must be designed in such a way that the publication region is larger than the subscription region (figure 4.1) and even though the publication regions overlap, the subscription region should not do so. The two regions actually play different roles for an airport depending on which side of the ownership exchange process the airport is in, whether the airport is the current owner of the attributes or the potential owner of them. The current owner publishes information about its aircraft while the aircraft is inside its publication region (region noted by the solid line), and it only releases the ownership of the aircraft attributes if the aircraft is outside its subscription region (region defined by the dashed line). Consequently, the potential owner receives information about the aircraft while the aircraft is inside its publication region, but it will only try to take over the ownership of the attributes when the aircraft has entered its subscription region. Since the subscription regions do not overlap, we will not have the problem of the oscillation effect.

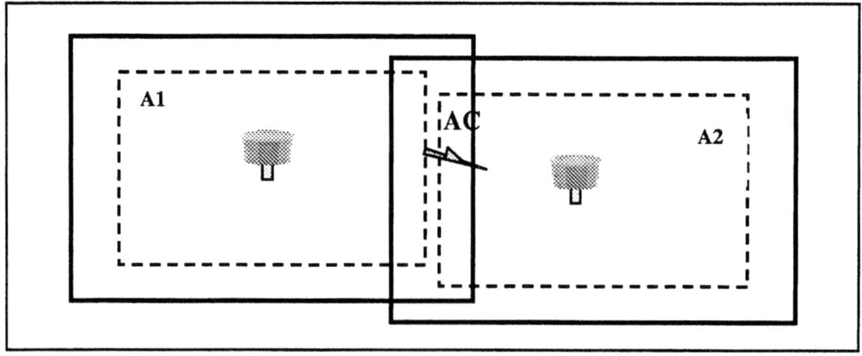

FIG. 4.1. *A simulation with two airports A1 and A2, where the dashed lines identify the subscription regions and the solid lines indicate the publication regions.*

This situation is illustrated in figure 4.1. In this scenario A2 will send a request to take over the ownership of the aircraft AC, when AC enters its subscription region. Later the request is granted and A2 becomes the new controller of AC. By that time AC is no longer within the subscription region of A1. Hence A1 will not send any request to take back the ownership of AC's attributes.

In the real system, each airport also has different airspace regions for controlling an aircraft. This can be thought of as a safety margin. Safety margin means that the airport can cover a large portion of the airspace and is able to control the aircraft, through radio communication, even if the aircraft is outside its radar range. But these measures are taken only in emergency situations, where for instance the airport that is supposed to take over the control, for some reasons, is not operating. Normally, airports relinquish the control of the aircraft long before the aircraft reaches the boundaries of that region. Hence, there is another much smaller region after which the airports exchange control of the aircraft.

After implementing the above solution we noticed an improvement in performance, which is logical since the number of ownership exchanges was decreased and therefore the unnecessary communication with the RTI was omitted.

4.1.2. Pending request. We mentioned the pending request problem in the previous sub-section and explained a situation where this problem could arise. But there are also other scenarios where one can be faced with this problem. For instance, the problem could also occur if we define a specific route for each flight. This means that each aircraft has to follow a certain route and that each controller airport knows who its successor is going to be for each flight. When different federates try to pull the same aircraft, the owner responds only to the predefined successor. This means that the other federates' requests would not be answered, and again we have some pending requests.

In the above scenario even if we do not define a specific route for each aircraft, we still could encounter the pending request problem if more than one federate tries to pull the ownership of the attributes. The reason is that in the RTI, the owner federate will (and can) only answer to one of the requests and the rest will be left unanswered.

Several solutions are now presented. The first solution was to employ a time-out mechanism. This means that if the requesting federate does not receive a response after a certain amount of time it would automatically send a cancellation and ask the RTI to discard its earlier request (time-out). This could be easily implemented, but the problem is to decide the exact time when the cancellation should be sent. If the cancellation is sent too soon or too late then we would experience the situation where we have attributes which do not have any owner, since no federate is taking over the ownership. The time-out approach could also be part of the RTI, where the pending requests would be timed out by the RTI after a certain amount of time. One solution would be to put the time-out period as a constant in the RTI.rid file. However, since the time needed to handle the request is application dependent (it may happen that in some applications it takes a longer time for a request to be answered, e.g. because of the latency in the network communication), the RTI should let the application programmer set this time.

Another solution would be to find another condition for cancelling the unanswered requests. In this solution each airport keeps track of the attribute ownership acquisition requests that it has sent to the RTI. For every pending request, at the same time as the federate receives the attribute reflection for the aircraft in question, it checks whether the aircraft is still within its subscription region.

Whenever the aircraft moves outside the region the airport sends a cancellation request to the RTI. This way if the RTI transfers the attribute ownership to the federate while the aircraft is still within the radar range, the federate can (and will) take over the ownership. Since the federate cancels its request as soon as the aircraft is outside its radar range then the RTI will never pass the ownership to the federate when the aircraft is outside the radar range. This method was tested and proved to be sufficient to solve the above problem.

4.2. Negotiated push. This approach looks more promising than the first one, since the federate which owns the aircraft attributes initiates the ownership transfer and controls the process. It invokes the "negotiatedAttributeOwnershipDivestiture" method whenever it is appropriate to do so (for instance when an aircraft moves out of its subscription region). Furthermore the attributes will never be left unowned since there must be a requesting federate before the process is completed. Of course, this is always the case since (as in the real system) the publication and subscription regions of the airports overlap each other.

In this approach, we did not have the problem with the oscillation effect. In the pull approach the exchange process could be completed even if the federates only had one radar range (same subscription and publication region). In the push approach there has to be a way for the federate to know when it is time to pass over the ownership of some attributes and still keep on publishing them until some other federate takes over the attributes in question, i.e. two different radar ranges are needed, and the subscription regions and the publication regions must have different sizes. But unlike the pull method here the subscription regions (regions noted by the dashed lines) of the adjacent airports have to overlap. The reason is that the controller airport will send an ownership divestiture request as soon as the aircraft is outside its subscription region, but (since subscription regions of different airports do not overlap) the aircraft would be outside the subscription region of the other airports as well. Hence, no other airport would answer to the ownership assumption request sent by the RTI and there would be a risk that the aircraft attributes will be left unowned. However the problem with the pending request still remained. Consider the scenario illustrated in figure 4.2 with four airports A1, A2, A3 and A4. For simplicity the regions for each airport are seen from above as squares. The squares with the solid lines indicate the publication regions and the ones with the dashed lines indicate the subscription regions.

The aircraft AC is flying from A1 towards A4. When AC leaves the subscription region of A1, it sends a "negotiatedAttributeOwnershipDivestiture" to the RTI, and the RTI sends a "requestAttributeOwnershipAssumption" to A2, A3 and A4. Since AC is within the subscription region of both A2 and A3, they will both send "attributeOwnershipAcquisition" to the RTI to acquire AC's ownership. Since only one federate is permitted to update the attributes, the RTI will send "attributeOwnershipAcquisitionNotification" to the federate whose request was last received. Assuming that the request which was answered belonged to

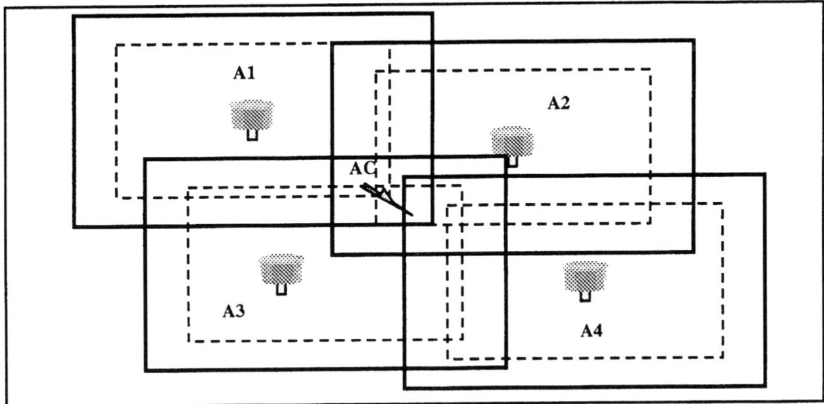

FIG. 4.2. *A simulation with four airports A1 to A4 and one aircraft (AC), where the dashed lines identify the subscription regions and the solid lines show the update regions of the airports.*

A2 then A3 has a pending request. This pending request may cause different problems. We have already discussed one of the problems in 4.1.1. Another consequence of the pending request problem is loss of aircraft, or more correctly put, unattended aircraft attributes. This situation, which is encountered in the push approach, occurs when A2, which is the new owner of the attributes, later decides to relinquish the ownership and sends a divestiture request to the RTI. However since there is already a pending ownership acquisition request, which was sent by A3, it can happen that the RTI passes the ownership directly to A3 without asking the other federates. In case the aircraft is already outside the subscription range of A3 (since A3 does not see the aircraft anymore), the attributes will be left unowned.

To solve this problem a method similar to the one in the pull approach can be used, and A3 was made responsible to regularly check the position of AC and to send a cancellation to the RTI as soon as AC approached the boundaries of its subscription region. This approach was shown to be sufficient in this case.

A potential problem with the push approach is that it may cause some unnecessary sending of information, especially when there are many aircraft in the air. A situation that must be avoided is the case where due to network latency neither A2 nor A3 respond on time, i.e. before AC has left their publication regions. Some problems with the latency in attribute ownership transfer and possible solutions for resolving them have been discussed by Roberts et al [16]. Therefore it is very important to choose the size of the publication and subscription regions correctly so that the intersection regions are sufficiently large and the aircraft do not leave the publication region before the ownership has been transferred. However, this is not a problem in the real system, since the publication and subscription regions are chosen quite carefully with large intersection regions.

Another problem with both approaches (push and pull) is that when there is more than one federate sending ownership acquisition requests, it is the RTI that decides the next owner of the attributes and there is no easy way for the current owner to choose its successor [17]. However, in an air traffic control scenario the current owner always chooses the next controller airport. The best solution for this problem would be to incorporate a mechanism in the RTI to allow federates to point out the new owner of their object attributes. For lack of such a mechanism, we were forced to find another solution to solve the problem. After some experimentation it was shown that the problem was much easier to solve in the push approach.

In some of the RTI methods and callbacks there is a neutral parameter, which is called "the tag". This parameter is of type string and acts as a container for passing information (which is not supported by the RTI) between federates. Often this parameter is passed as NULL, but in our case the controller federate uses this tag to send the name of its successor via the ownership divestiture request. As described earlier all federates which are eligible to take over the ownership of the attributes will receive an ownership assumption request from the RTI. It means that they will also receive whatever information is stored inside the tag parameter. This way only the federate that has the same name as the name passed via the tag will send an ownership acquisition request and will take over the ownership of the attributes. This solution was easy to implement and worked satisfactorily. Implementing a similar solution for the pull approach was much more complicated.

5. Performance analysis. This section describes the results of our experiments, which were conducted to investigate the performance and analyse the two ownership transfer approaches described in section 4. The experiments reported in this paper have been performed on an MPP- the Fujitsu AP3000 with 32 nodes, running on UNIX(r) System V, Release 4.0. The platform has been developed in C++, and the RTI, which has been used in our experimental federation, is DMSO RTI-1.3v5.

Measuring the performance of a system and especially the execution time is not a trivial problem. First of all if the system load is not consistent, the same program will have different execution times each time it is run. Another problem is that since our simulations involve several federates, the order in which federates are executed may affect the execution time and the number of messages communicated with the RTI.

We have taken the following measures to solve these problems:

1. We start measuring the time after the initialisation period has completed and when the simulation has reached a steady state. The time measurement is finished before the simulation has reached its end. This way we avoid the inconsistency during the starting and the ending stages of the simulation and ensure that during the time interval of our measurement, all federates are running in a stable state.

2. For each set of parameters we run the simulation 5 to 10 times and

then take the average time and number of messages. This is done to reduce (and compensate for) the effects caused by the random number generators and the execution order of federates.

5.1. Oscillation effect. The problem with the oscillation is dependent on the size of the intersection area between the publication/subscription regions of two adjacent airports and the speed of the aircraft flying from one to another. Table 5.1 shows the results of our experiments for two airports with different intersection region sizes, and different aircraft speeds. The oscillation frequency is the average number of time the aircraft changed ownership, while they were inside the intersection area.

TABLE 5.1
The relation between the oscillation frequency, the speed of the aircraft and the size of the intersection area

No.of airports	No.of aircraft	Ave Speed (km/s)	Width of intersection area (km)	ave no.of oscillations
2	10	20	200	6
2	10	20	1200	39
2	10	20	2200	56
2	10	40	200	3
2	10	40	1200	19
2	10	40	2200	27

As we can see, from the results the oscillation frequency is proportional to the size of the intersection area and inversely proportional to the speed of the aircraft. The longer the aircraft stays within this area the more often the oscillation occurs and when the speed is doubled, the oscillation frequency is almost cut to half. Additional experiments with variable number of airports and aircraft showed similar results and the same pattern of behaviour was confirmed.

5.2. Pending requests. The frequency and the number of the pending requests is proportional to the number of the airports, location of the airports with respect to each other, number of the aircraft in the simulation and the simulation time. The location of the airports is important because in our simulation it affects the flying route of an aircraft and whether the aircraft crosses the boundaries of more than one airport as it flies out of the controller's airspace.

In Table 5.2 we present part of our experiments with different number of airports (in this case with 2, 4, 6, 9) and two different numbers of aircraft (5 and 10) for each airport. The simulation time in each case is 1000 time units, but the sampled time is 800 time units.

As expected, we can see from Table 5.2 that the number of pending requests is very much dependent on the location of the airports. In the case of the two airports we do not experience any pending requests, since there is never more than one airport trying to take over the ownership. In the case of 6 airports, the number of pending requests is less than when we have 4 or 9 airports. The reason for this is that when we experiment with 4 and 9 airports, the airports

TABLE 5.2
Number of pending requests with respect to the number of airports, aircraft and the total number of ownership exchanges

No.of airports	No.of aircraft/airport	Total no. of pending requests	Total no.of ownership exchanges	No. of Unattended aircraft
2	5	0	20	0
2	10	0	38	0
4	5	8	35	3
4	10	18	66	6
6	5	4	67	2
6	10	9	144	4
9	5	11	89	5
9	10	25	214	11

are located in a way that they form a quadratic area. This means that the probability that an aircraft enters the regions of the space, which is shared by more than two airports is much higher compared to case of 6 airports (figure 5.1).

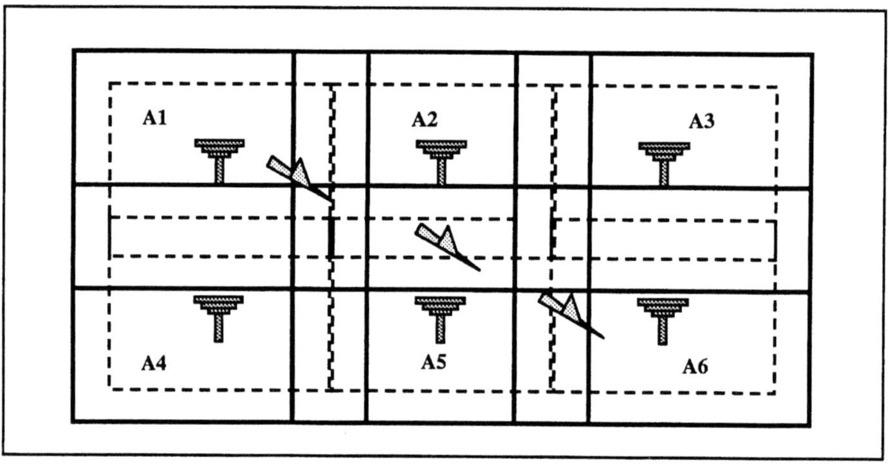

FIG. 5.1. *A simulation with six airports A1 to A6, where all the aircraft moving from A1 towards A6 will cross the boundaries of one airport at a time.*

5.3. Unattended aircraft. As mentioned earlier, pending requests create a situation where some aircraft lose contact with their controller airports, and since no airport is taking over the control of the aircraft, their attributes will be unowned.

The results presented in table 5.2 indicate that the number of pending requests is proportional to the number of the unattended aircraft (or aircraft attributes) and the more we run the simulation the more the aircraft lose their owners. In table 5.2, we have presented the number of unattended aircraft after 1000 time steps.

5.4. Negotiated push and pull. Table 5.3 presents a comparison between the *pull* and the *negotiated push* methods with regard to the execution time and the number of messages communicated between federates and the RTI in each case. The experiments have been conducted with different number of airports and aircraft.

The aim of these experiments is to show which method is faster and also to show the communications involved in each method, since this could affect the total system performance and latency, if for instance, the communication cost is increased.

TABLE 5.3
A comparison between push and pull methods with regard to the execution time and the number of messages communicated with RTI

No.of airports	No.of aircraft/airport	Time in seconds		Communications with RTI	
		push	pull	push	pull
6	6	103	108	321	305
6	10	235	244	512	486
9	6	158	172	510	464
9	10	230	247	805	740

The RTI communication, which is mentioned in the above table, is only referring to the communication, which involves ownership transfer of attributes. We assume that the remaining part of the communication, e.g. attribute updates and reflections, are the same in both approaches. Furthermore, the transfer scenarios and the number of ownership exchanges in both push and pull cases were the same.

As we can see, the push method is slightly faster than the corresponding pull method. One explanation could be that the pull involves more computation, since each federate has to check all remote objects, (aircraft that do not belong to the airport) at each time step to see whether any of them has entered its airspace. This computation cost increases as more aircraft are simulated.

At the same time federates communicate more with the RTI in the push method. This could be important if the cost of sending messages between federates and the RTI is increased, e.g. if the nodes that federates and the RTI are running on, are (geographically) located further apart.

6. Lessons learned. In this section we summarize our observations and experiences with working with a distributed federated environment, such as ATC, and the RTI. We also compare the two ownership exchange methods described in this paper, namely pull and negotiated push and list out some of the RTI problems we noticed while running our federation.

6.1. ATC. The air traffic control simulation that we implemented could have been modelled and developed as a single monotonic simulation program, where all the airports and aircraft were simulated by the same simulation model. There would be some advantages in having one simulation instead of a set of simulators connected via the RTI. For instance having one simulation would perhaps eliminate many of the problems we experienced with ownership exchange and in some cases would make it run faster, especially if it is a parallel simulation using multiple CPUs simultaneously as discussed in the paper by Wieland [18]. But it all depends on the objectives of the simulation and the reason why the simulators have been developed. Having a distributed simulation would most definitely make the simulation more scalable. It will also make it suitable for training purposes, where air traffic controllers can actually be located at different geographical locations and participate in a joint simulation.

Another reason would be the reusability of the software. Having created an airport federate, one could simulate several federates with different characteristics and run them in the same simulation. Airport federates could also be reused in federations, where they interact with other (and different) kinds of federates.

One obvious reason for simulating a distributed ATC, as in our case, would be the possibility that it provides for investigating the services provided by the RTI.

But working with the RTI in some situations has proven to be less than satisfactory. The system was not always stable especially when a large number of airports and aircraft were simulated and lots of messages were exchanged. We also experienced problems with the Ownership Management services, e.g. in some cases the RTI did not deliver messages as was expected. These problems will be discussed later in section 6.3.

6.2. Comparison of push and pull methods. As explained in section 3, in our first implementation of ownership exchange between federates we used the pull method. This method as described earlier demonstrated some problems and we noticed the oscillation effect. Aircraft started to oscillate between different owners in the intersection zone.

Our first solution was to stop this at the owner side by introducing some restrictions. For instance, if the owner was also the destination airport or if the aircraft was moving toward the airport then no response was sent back to the RTI approving release of ownership. But then the problem with the pending request was encountered, which caused trouble at the requesting federates. Unfortunately there is no method in the RTI through which the requesting federate could be notified that its request had not been granted. In order to solve this problem we used a method which we called, "conditional cancellation". This means that each federate keeps track of all attribute acquisition requests it sends out and for each unanswered request it checks whether the aircraft in question is still within its subscription range, as discussed in section 4.1.2. If the aircraft has moved out of the region a request is sent to the RTI to cancel the corresponding acquisition request.

We also introduced restrictions at the requesting federates' side. No request for ownership acquisition would be sent if the aircraft was flying away from the airport. This was done by some geometrical calculations, in a way that if the angle between the direction vector of the aircraft and the vector drawn from aircraft to the airport was less than 90 degrees then one could assume that the aircraft is moving towards the airport. If the angle was more than 90 degrees then the aircraft was moving away from the airport.

Unfortunately this condition is not sufficient. Depending on how the airports are located, one could think of cases where an aircraft has to pass through the territory of an airport even though it is not moving towards it. The final solution was to define the publication and subscription regions in a way that the publication regions were larger than the subscription regions and the subscription regions did not overlap. In this way a federate would send a request to take over the control of an aircraft if it is inside its subscription region. Since the subscription regions do not overlap no other federate will try to take over the control of the aircraft, while it is inside the subscription region of the controller federate.

Our next implementation was the negotiated push. This method turned out to be much better even though it was not our first choice. There are a couple of things that we had to look out for when we used this approach. First of all the regions had to be chosen carefully and that the subscription and publication regions should have two different sizes. Second, in the push method the owner cannot easily choose its successor, which would be desirable in scenarios like ours. Unfortunately this problem exists in the pull approach as well. But we noticed that it is much easier to solve this problem in the push method by utilizing the tag parameter in the RTI ambassador methods. By passing the name of the next owner via the ownership divestiture method, the current owner could inform the other federates eligible to take over the ownership, which federate is the desired successor. The same solution for the pull method would be more complicated and require more computation at the side of the potential owners.

The number of aircraft can also be important in the push approach, since there is more communication involved in this method compared to the previous one.

The problem with the pending request also occurred in this approach. In this respect the two methods demonstrated the same problem and the solution for resolving this was the same, i.e. the conditional cancellation method.

Another reason why we consider the push method to be superior to the pull method is because of the fact that the amount of time spent by each federate on checking the conditions for pull is usually greater than the corresponding time spent for push. In the first method each federate has to check all the external (remote) objects (which it can see) at each time step to check whether any of the objects has entered its airspace, and then send a request to acquire the ownership of that particular object (pulling the ownership). But in the push approach each airport is only concerned about its own objects and does not do anything until one of its aircraft crosses the subscription region of the airport. When this event occurs, the airport tries to relinquish the ownership of the aircraft by notifying the RTI (pushing the ownership). Meanwhile the other federates are not concerned with the movement of objects outside their airspace. Since in our simulator the number of external objects (that can be seen by each airport) is usually greater than the number of the local objects, there will be less time spent on checking the conditions in push compared to pull. One could say that pull is somewhat time-driven as oppose to push which is more event-driven, using the analogy used in [19].

6.3. Problems with the RTI. While working with ATC, we experienced two kinds of problems with the RTI, the insufficient and incomplete three-way handshaking in OWM services and run-time problems.

The three-way handshaking and the pending request problems were discussed in section 4.1.2. As mentioned, our observations indicated that OWM services in the RTI are missing the proper "three-way handshake" procedure that is needed to provide good assurance of successful transfer of ownership between distributed asynchronous processes. Here we present a method on how to complete this procedure.

Our suggestion is that there should exist a method in the RTI-ambassador that allows federates to reject an ownership release request from the RTI, which is a result of an ownership acquisition request. If this method exists and it is invoked, there are two alternative ways that the RTI could react.

1. the RTI cancels the corresponding acquisition request and then informs the federate that its request has been cancelled.

2. the RTI informs the federate in question that its request has been rejected and let the federate decide whether it wants to cancel the request or not.

There are advantages and disadvantages to both methods, and it depends on the underlying simulated system to decide which one is more efficient. However in our model the second alternative would eliminate unnecessary communication, and is more efficient and desirable.

We also encountered some run-time problems while running our simulation. There are two types of problems and both occurred when we experimented with large numbers of federates and objects. In the first case we noticed that every time we simulated a large number of objects and large numbers of messages were communicated with the RTI, errors like bus error would stop federates from running. This problem occurred on a consistent basis, regardless of the method (push or pull).

In the second case, which occurred when OWM services were stressed, and many messages were being communicated, the RTI did not respond as expected. The following cases demonstrate situations where this happened.

1. In the push method it happened several times when many acquisition requests were invoked by the federates, the RTI did not send the acquisition notification to any federate, but it assumed that the ownership had been transferred and sent a divestiture notification to the owner.

2. In another case the RTI did not send any ownership assumption to the concerned federates, even though an ownership divestiture was initiated by the owner, and still it raised error when the owner kept publishing the attributes. This problem was encountered when we were experimenting with the

push method.

3. In one particular case when some attributes had changed ownership three times, the RTI complained every time the latest owner published the attributes, even though it had given the ownership to the federate.

Based on our experiments, we believe that the above problems were caused by the version of the RTI that we were using. We believe that this version is not robust enough to handle and facilitate large number of simultaneous ownership transfers.

7. Conclusion. In this paper we have summarized our experiences in the use of the ownership management services of the DMSO/RTI 1.3 in the context of a distributed air traffic control simulator. We discussed some ownership management problems, the oscillation effect and the pending request, that can arise and our approaches to resolving these problems (e.g. different regions and conditional cancellation). Generally, the former problem is caused by federates having overlapping subscription regions that wish to transfer object attributes between them. The latter problem is due to the lack of a proper general three-way handshaking mechanism in the RTI, which will allow a federate to reject any request for an acquisition of its object attributes.

We have given a detailed comparison of two different methods for transferring attribute ownership among federates, the pull and the negotiated push method. Our experiences indicate that in scenarios like ours, the push method is more efficient and accurate. In both cases, the choice of the subscription and publication regions is crucial. In the push approach, both publication and subscription regions should overlap, but in the case of the pull, the subscription regions should not overlap.

We have shown that the push method performs faster than the pull method, although it involves more communication with the RTI. Communication cost is important, especially when the number of airports and aircraft in the federation are increased. But according to our experiments the push method continues to be faster than pull, even when we increased the number of objects. This is due to the fact that in the pull approach the number of objects is proportional to the computation time (the time spent by a federate to check whether any remote object has entered its airspace). The more objects we simulate the more time the federate has to spend on checking the above condition.

We have also detected a shortcoming in the current RTI interface specification, namely the lack of a proper three-way handshaking procedure in the Ownership Management services. The reason is that there is no mechanism implemented in the RTI that allows a federate to reject an attribute release request. So when a federate does not get an answer to its request it does not know whether the request has been rejected or the other federate is just taking its time. This leads to the pending request problem, which in turn results in unattended aircraft attributes (or loss of aircraft). In general, this problem will arise whenever several federates with overlapping subscription regions try to acquire the attributes of an owner federate. This problem should be addressed since it will lead to erroneous and invalid simulations, as in the case of our unattended aircraft of the ATC system. The best solution for resolving this problem would be to enhance the current RTI by adding a method for rejecting attribute release requests. We have suggested two alternatives on how the RTI should react to an invocation of the suggested method. Work is currently in progress to test the efficacy of these solutions.

It has also been shown that although in some cases the current owner of attributes should be the one that chooses its successor, with the current the RTI implementation, it is the RTI that makes the decision of who the next owner should be.

REFERENCES

[1] J. S. Dahmann, R. M. Fujimoto and R. M. Weatherly, *The Department of Defense High Level Architecture*, in Proceedings of the 1997 Winter Simulation Conference, 1997.

[2] Judith S. Dahmann, *High Level Architecture for Simulation*, in Proceedings of the First International Workshop on Distributed Interactive Simulation and Real-Time Application, 1997.

[3] Judith S. Dahmann, R. Fujimoto, R.M.Weatherly, *The DoD HLA: An Update*, in Proceedings of 1998 Winter Simulation Conference, Washington D.C., Dec. 1998, pp. 797–804.

[4] Daniel J. Van Hook, Steven J. Rak, James O. Calvin, *Approaches to RTI Implementation of HLA Data Distribution Management Services*, the 15th Workshop on Standards for the Interoperability of Distributed Simulations, September 1996.

[5] A. Boukerche and A. Roy, *A Dynamic Grid-Based Multicast Algorithm for Data Distribution Management*, in Proceedings of the 4th IEEE Distributed Simulation and Real-Time Applications, San Francisco, USA, August 2000, pp. 47–54.

[6] G. Tan, Y. S. Zhang, R. Ayani and F. Moradi, *A Hybrid Approach to Data Distribution Management*, in Proceedings of 4th IEEE Distributed Simulation and Real-Time Applications, San Francisco, USA, August 2000, pp. 55–61.

[7] R. M. Fujimoto and R. M. Weatherly, *Time Management in the DoD HLA*, proceedings of 1996 Parallel and Distributed Simulation Workshop, PADS'96, May 1996.

[8] *HLA Interface Specification, Version 1.3* (April 02, 1998) (http://hla.dmso.mil/hla/tech/ifspec/).

[9] Anita Adams Zabek, Annette Wilson and Dr Mary (Connie) Fischer, *The ALSP Joint Training Confederation and the DOD High Level Architecture*, Proceedings of the 14th DIS Workshop, March. 1996.

[10] Nico Kuijpers, Johan Lukkien, Bas Huijbrechts and Marco Brasse, *Applying Data Distribution Management and Ownership Management Services of the HLA Interface Specification*, in Proceedings of the Simulation Interoperability Workshop, Fall 1999.

[11] Zhian Li, Charles M. Macal and Michael R. Nevins, *The Problem of Object Ownership Transfer in HLA-Compliant Logistics Simulations*, in Proceedings of the Simulation Interoperability Workshop, Fall 1998.

[12] Zhian Li, Charles M. Macal and Michael R. Nevins, *Ownership Transfer of Non-Federate Object and Time Management in Developing HLA Compliant Logistics Model*, in Proceedings of the Simulation Interoperability Workshop, Spring 1998.

[13] *HLA Data Distribution Management: Design Documents Version 0.7* (November 12, 1997) (http://hla.dmso.mil/hla/tech/ifspec/).

[14] *HLA Object Management: Design Documents Version 1.0* (August 15, 1996) (http://hla.dmso.mil/hla/tech/ifspec/).

[15] *Aeronautical Information Publication*, Civil Aviation Authority of Singapore, Spring 1999.

[16] DAVID J. ROBERTS, ANDY T. RICHARDSON AND PAUL M. SHARKEY, *Optimising Exchange of Attribute Ownership in the DMSO RTI*, in Proceedings of the Simulation Interoperability Workshop, Spring 1998.

[17] BRET R. GIVENS, CINDY MARTIN AND DAVID O'QUINN, *Using HLA Ownership Management to Migrate Legacy Applications*, in Proceedings of the Simulation Interoperability Workshop, Fall 1998.

[18] FREDERICK WIELAND AND DAVID JEFFERSON, *Case Studies in Serial and Parallel Simulation*, in Proceedings of the International Conference on Parallel Processing, August 1989, pp. III-255–III-258.

[19] FREDERICK WIELAND, *Parallel Simulation for Aviation Applications*, in Proceedings of the 1998 Winter Simulation Conference, December 1998, pp. 1191–1198.

[20] FARSHAD MORADI, RASSUL AYANI AND GARY TAN, *Object and Ownership Management in Air Traffic Control Simulation*, Proceedings of IEEE DIS-RT99, Maryland, U.S.A., October 1999, pp. 41–48.

COMMUNICATION OVERHEAD ON DISTRIBUTED MEMORY MACHINES

SHIPING CHEN* AND JINGLING XUE†

Abstract. Distributed memory machines (DMMs) can provide scalable and flexible high performance if the relatively high startup overhead on these machines can be amortised. This paper studies communication overhead on DMMs. Based on *LogP*, a communication cost model is developed to quantify separately the effects of send, receive and network contention on overall communication overhead. This cost model is applied to estimate communication overhead for four commercial multicomputer systems. Two observations for send and receive overheads on DMMs are discussed.

Key words. Distributed Memory Machines, Communication Overhead, Cost Model

1. Introduction. Distributed memory machines (DMMs) promise to provide scalable and flexible high performance [10, 16]. Because each processor has its own local memory, DMMs are more scalable than shared memory machines. With the advances in static and dynamic network technologies, a large number of processors can be easily configured into various topologies at relatively low cost. For example, a CM-5 machine can contain up to 16,384 processors connected under different topologies [11]. However, communication overhead on DMMs is relatively high. Without a global address space, communication between processors must be explicitly expressed in programs for remote data access. Even on modern DMMs, a remote data access still takes one or two orders of magnitude longer than a local data access.

The expensive communication has a significant effect on computation partitioning on DMMs [1, 15]. While small task sizes are expected to exploit a lot of parallelism, large task sizes can reduce communication frequency and overall communication cost. Therefore, accurate communication cost models are essential to estimate the performance of a given partitioning scheme on a specific distributed memory machine.

In this paper, we study communication overhead on DMMs. First, a communication cost model is proposed. This cost model considers overheads for send, receive and network contention separately, enabling us to isolate their effects on overall communication overhead. Second, we discuss methods that can be used to determine by experiments the values of the communication parameters in our cost model for a given distributed memory machine. The four concrete communication cost models are provided for IBM SP2, TMC CM-5, FUJITSU AP1000 and a network of DEC Alpha workstations. Finally, the problem concerning the large overhead of the first send call in a processor and the problem concerning blocking and non-blocking send calls are investigated. A simple method is

*CSIRO Mathematical and Information Science, Australia (shiping.chen@cmis.csiro.au).

†School of Computer Science and Engineering, The University of New South Wales, Sydney NSW 2052, Australia (jxue@cse.unsw.edu.au).

provided to reduce the effect of the first send call in a processor for a class of data-parallel programs.

The rest of this paper is organized as follows. Section 2 presents our communication cost model. Section 3 describes the concrete communication cost models for four multicomputer systems. Two observations for communication overhead on DMMs are elaborated in Section 4. Section 5 discusses the related work. Section 6 concludes.

2. Modeling Communication Overheads. Our communication cost model is based on the *LogP* model reported in [5, 17]. We model point-to-point communication overhead on a distributed memory machine at application level.

2.1. *LogP* Abstract Machine. In the *LogP* model, a multicomputer consists of a set of processors connected via an interconnection network. The processors can communicate with each other via point-to-point message passing. Such a *LogP* abstract machine is characterized in terms of four parameters, as shown in Figure 2.1:

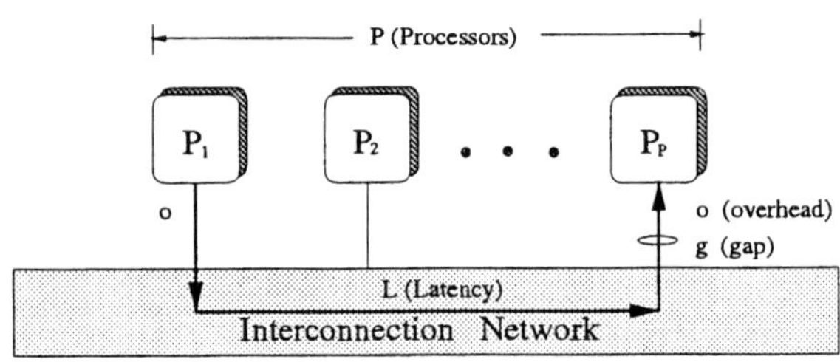

FIG. 2.1. *LogP abstract machine*

L: the *latency*, defined as the time needed to transfer a message containing a word from a source processor to a target processor.

o: the *overhead*, defined as the length of time that a processor is engaged in the transmission or reception of a message.

g: the *gap*, defined as the minimum time interval between consecutive message transmissions or receptions at a processor.

P: the number of processors that communicate simultaneously in the system.

Figure 2.2 shows how the *LogP* model can be used to analyse the communication overheads for four commonly used communication modes.

In the standard mode, the sender P_1 sends a message to the receiver P_2 directly, as shown in Figure 2.2 (a). First, P_1 spends o time units on injecting the message into the network. Then, it takes L time units for the message to reach its destination P_2 via the network (switches and routers). Finally, P_2 spends another o time units on copying the message from the system space to its local space and performing some other operations such as message unpacking

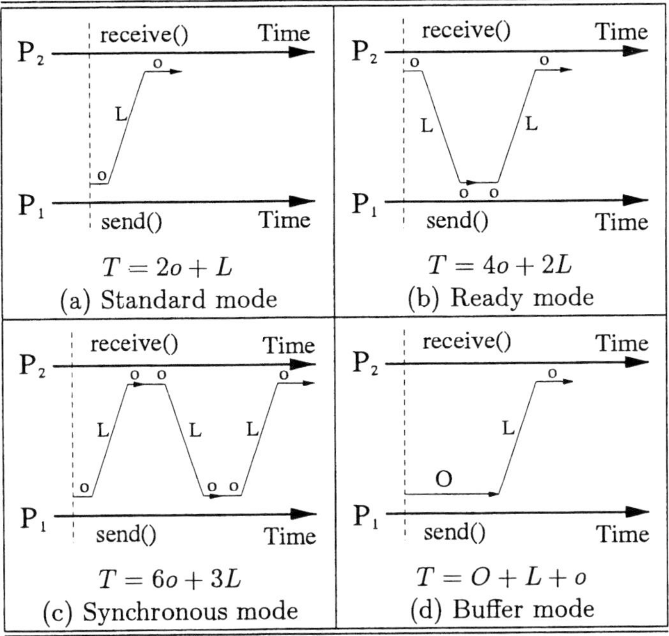

FIG. 2.2. *Four communication modes in the LogP model.*

and checksum validation. Therefore, the total message communication time in the standard mode is $T = 2o + L$.

The other three communication modes can be modeled similarly. In the ready mode, as shown in Figure 2.2 (b), the send call on P_1 is blocked until a ready acknowledgement from the matching receive call on P_2 is received. Thus, the total cost is $T = 4o + 2L$. In the synchronous mode illustrated in Figure 2.2 (c), the two communicating processors need to handshake with each other via two pairs of send and receive calls before the message is transmitted. Thus, the total cost is $T = 6o + 3L$. The buffer mode, as shown in Figure 2.2 (d), is similar to the standard mode except that the sender P_1 may pay extra costs in time and memory for buffering the data to be sent. The performance of the buffer mode depends on specific hardware architecture and software implementation (hidden in O). The total cost is thus $T = O + L + o$.

In machines where the overhead o dominates the gap g, g can be ignored [5].

The message communication time formulas given in Figure 2.2 reflect the absolute time cost for one communication rather than the time costs charged to individual processors. Therefore, these formulas are difficult to use in applications requiring an approximation of the communication overhead on either the sender or the receiver. Furthermore, the *LogP* parameters, namely, the latency (L) and the overhead (o), in a real machine are nontrivial to measure accurately.

2.2. Our Communication Cost Model. To analyse the communication overheads charged to individual processors, the effects of send and receive on overall communication overhead are quantified separately at application level. As a result, our communication cost model provides a layer of abstraction to encapsulate the complexity of overheads, latency and communication modes as shown in Figure 2.3.

In this work, only point-to-point blocking communication is considered. The development of our cost model is based on the following two definitions.

DEFINITION 2.1 (Send Overhead). *Assume that the two matching communication calls,* send() *and* receive(), *start at the same time* t_1 *on two processors* P_1 *and* P_2, *respectively. If* send() *exits at* t_2, *the* send overhead *on* P_1 *is* $T_{send} = t_2 - t_1$.

DEFINITION 2.2 (Receive Overhead). *Assume that the two matching communication calls,* send() *and* receive(), *start at the same time* t_1 *on two processors* P_1 *and* P_2, *respectively. If* receive() *exits at* t_2, *the* receive overhead *on* P_2 *is* $T_{recv} = t_2 - t_1$.

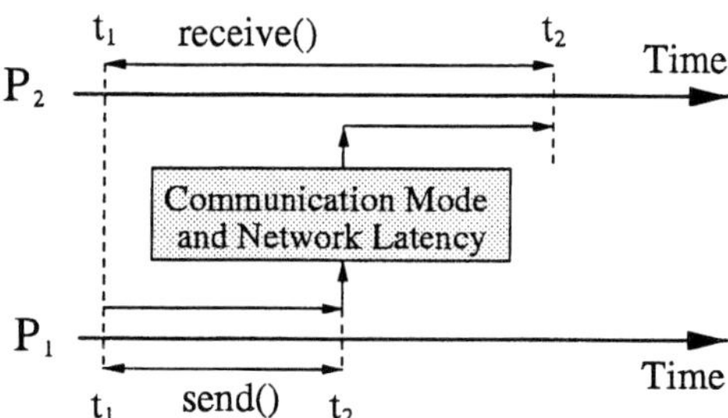

FIG. 2.3. *Our communication cost model for send and receive*

The send and receive overheads, illustrated in Figure 2.3, are estimated by:

$$\begin{cases} T_s = L_s + g_s B \\ T_r = L_r + g_r B \end{cases} \tag{2.1}$$

where L_s (L_r) is called the *startup overhead* for send (receive), g_s (g_r) is the time spent on sending (receiving) an unit message, and B is the size of a message.

Another communication overhead is *network contention.* For example, workstations on bus-based networks share physical media, such as coax or fiber optic. Collisions may occur when more than one node attempt to send messages at the same time. Intuitively, the more nodes involved in simultaneous message transmissions or receptions, the higher possibility that network contention will occur. If the impact of network contention is considered, the send overhead can be estimated as follows:

$$T_s = L_s + g_s B + \gamma P \tag{2.2}$$

where γ is the network contention rate per processor in a P-processor machine.

3. Measuring Communication Parameters.

3.1. Methods. Both Ping and Ping-Pang methods are used to determine the the communication parameters in our cost model.

(a) Ping	(b) Ping-Pang

<pre>T = 0
if pid = P1 then
   do k = 1, N
      sleep(t)
      t1 = timer()
      send(message)
      t2 = timer()
      T = T + (t2 - t1)
   enddo
   Ts = T/N
else if pid = P2 then
   do k = 1, N
      receive(message)
   enddo
endif</pre>
	<pre>T = 0
	if pid = P1 then
	   do k = 1, N
	      t1 = timer()
	      send(message)
	      receive(message)
	      t2 = timer()
	      T = T + (t2 - t1)/2
	   enddo
	   Tr = T/N
	else if pid = P2 then
	   do k = 1, N
	      receive(message)
	      send(message)
	   enddo
	endif</pre>

FIG. 3.1. *The SPMD code for Ping and Ping-Pang experiments*

The *Ping* method [2] is chosen to measure the send overhead, which is the elapsed time of send() on the sending processor. The elapsed time may or may not include network latency and synchronisation cost depending on the communication modes used. In this way, different communication modes are encapsulated in our cost model from the viewpoint of the sending side. The algorithm used in the Ping experiment is given in Figure 3.1 (a) and illustrated in Figure 3.2 (a). Note that the Ping experiment is designed so that (1) the common pattern of parallel computing, i.e. communication-computation-communication, is reflected in the cost model, and (2) the stochastic factors resulting from various communication mechanisms used on different parallel machines such as interruptions and handshaking tend are minimised.

Due to the lack of a global clock across the processors in a distributed memory machine, it is more difficult to measure accurately the receive overhead. Therefore, the *Ping-Pang* method, as depicted in Figure 3.2 (b) and illustrated in Figure 3.2 (b), is used. P_1 sends a message to P_2. As soon as P_2 receives the message, P_2 sends the same message back to P_1 immediately. The whole process costs approximately twice the receive overhead, i.e. $T = t_2 - t_1 = 2T_r$. Thus, the receive overhead is $T_r = \frac{t_2 - t_1}{2}$.

To quantify the effect of network contention on overall communication overhead, we vary the number of processors used in our experiments to simulate the network contention. By fitting the experimental data thus obtained into

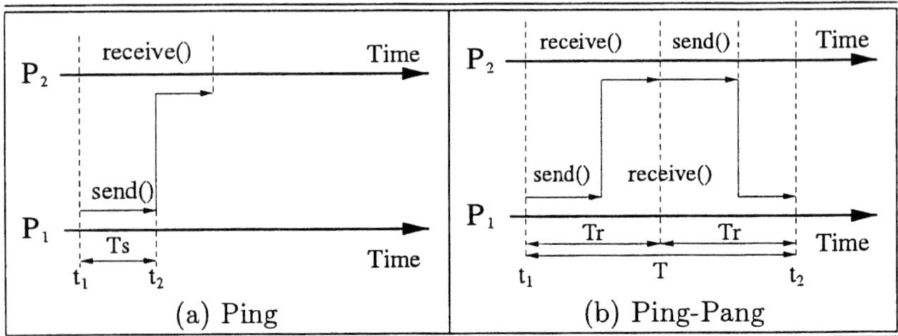

FIG. 3.2. *Ping and ping-pang experiments.*

Equation 2.2, the network contention rate γ in the formula can be derived.

3.2. Machines. Our experiments were performed on the following four commercial multicomputer systems:

- IBM SP2,
- TMC CM-5,
- FUJITSU AP1000, and
- A network of DEC Alpha workstations (NOWs).

The general information about these machines is given in Table 3.1.[1]

System	Processor	Network	Software	Affiliation
IBM SP2	RS6000	HPS	MPI	QPSF
TMC CM-5	Sparc	DN/CN	CMMD	SACPC
FUJITSU AP1000	Sparc	T-net	CAP-II	ANUSF
DEC NOWs	Alpha	Ethernet	PVM	UNE

TABLE 3.1
Four multicomputer systems

The SP2 comprises 22 *nodes*, which are configured into three *frames* connected by High Performance Switch (HPS) network. Each node has an RS6000 processor with 128MB memory and a 2GB internal disk drive. The 22 nodes are divided into two groups: 8 *thin* nodes deliver 230 MFLOPS each and 14 *wide* nodes deliver about 266 MFLOPS each. Two protocols are supported: *IP* protocol and *US* protocol [12]. The communication software system used on SP2 is MPI (Message Passing Interface) [21].

The CM-5 has 128 processor nodes. Each node has a Sparc processor with 8MB memory. These processors are controlled by a 150 MHz HyperSparc Sparc-Server 20 with 128Mb memory. The CM-5 uses two different networks, called *Control Network* (CN) and *Data Network* (DN), to transfer instructions and

[1]QPSF — Queensland Parallel Supercomputing Foundation; SACPC — South Australian Center for Parallel Computing; ANUSF — Australian National University Supercomputer Faculty; and UNE — University of New England.

data, respectively. The communication software used in our experiments is its native message-passing development toolkit called *CMMD* [19].

The AP1000 consists of 128 nodes, called *cells*, each of which has a Sparc processor running at 25MHz with 16MB memory. All processors are connected to three networks called the torus network (T-net), broadcast network (B-net) and synchronization network (S-net). The T-net provides the point-to-point communication using the wormhole routing with a 25MB/sec of bandwidth. The native communication library *CAP-II* is used for message passing [8].

Our network of workstations includes 10 DEC Alpha workstations running at 150MHz with 128MB memory. The workstations are connected by an Ethernet network. The communication software used is PVM (Parallel Virtual Machine) [9].

3.3. Experiments. Experiments using different data sizes were repeated to measure the communication overheads on four multicomputer systems. The parameters in our cost model are obtained using the linear regression tool *Splus* [3].

Little network contentions are observed on SP2, CM-5 and AP1000. The communication overheads on these machines can be approximated using Equation 2.1. The experimental data and the send and receive cost formulas derived for these three machines are shown in Figure 3.3 and further compared in Figure 3.4.

The SP2 with the US protocol has the lowest send and receive overheads among the four configurations displayed in Figure 3.3. This is because the US protocol can make the best use of *High-Performance Adapter-2* and *Resource Manager* [12]. On the other hand, the communication using the IP protocol suffers the highest startup overhead for both send and receive. This shows that network protocols do play an important role in communication performance. Due to the wide bandwidth on SP2, the send and receive overheads in these two cases are not sensitive to message sizes.

The CM-5 has a relatively higher send overhead because the communication mode used its message passing software (CMMD) is synchronous. Before sending a message from P_1 to P_2, P_1 first sends P_2 a *request to send* (RTS) message and is then blocked until it has received an *acknowledgement* (ACK) from P_2. Then P_1 transmits the message until the message transmission is complete. Despite of the relatively higher send overhead, two benefits make the synchronous mode on CM-5 worthwhile. First, the synchronous communication provides safe data transmissions between processors. Second, the synchronous communication can enhance the communication performance by reducing some *control overhead* (e.g., interrupt overhead) on the receiving side. This is confirmed by the CM-5 receive cost formula in Figure 3.3 (f).

The AP1000 has a huge difference between its send and receive overheads. This is because AP1000 uses the standard communication mode and has a narrow bandwidth.

Due to network contention, Equation 2.2 is used in our network of DEC

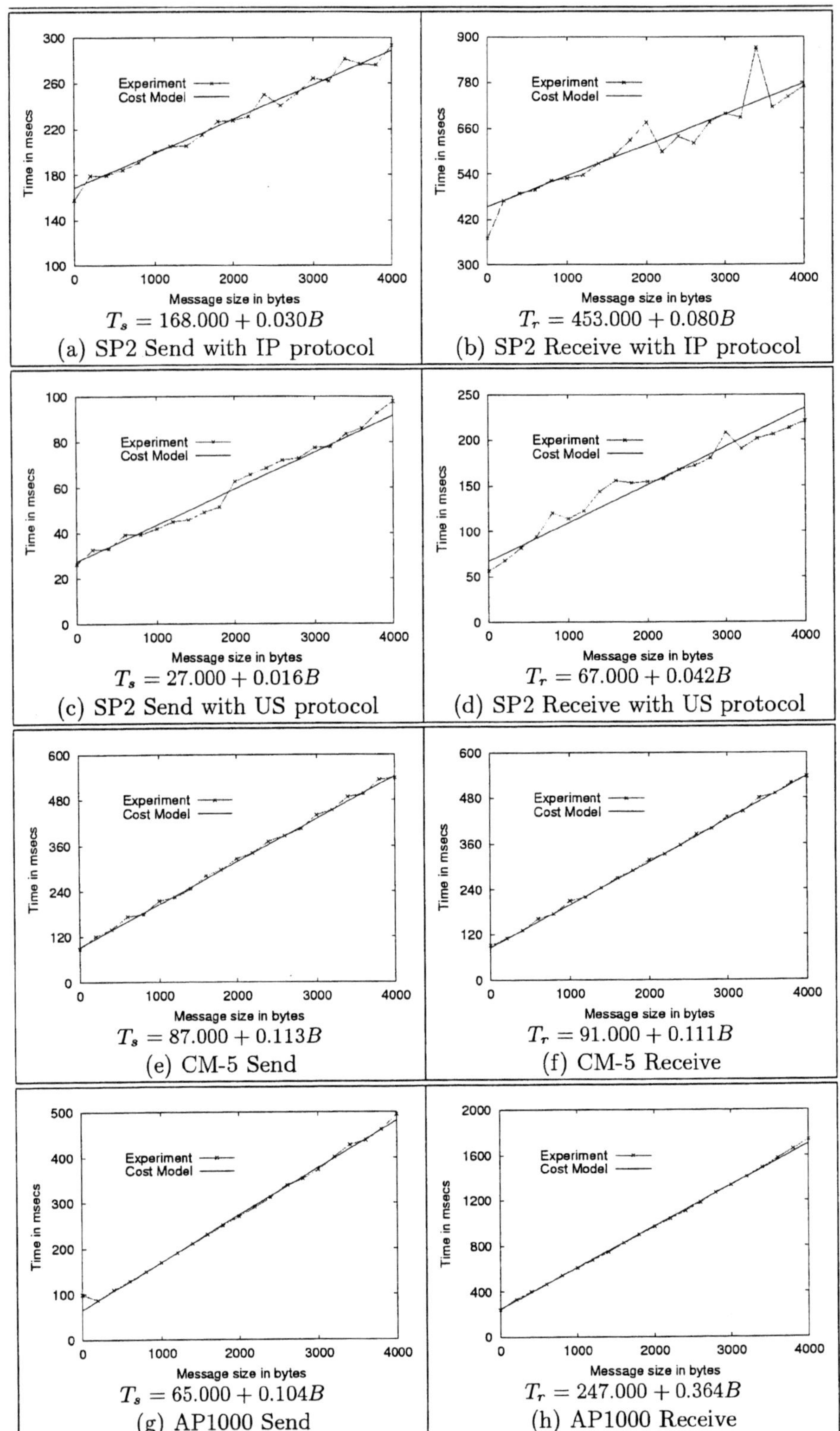

FIG. 3.3. *Communication cost models for SP2, CM-5 and AP1000 (μsecs)*

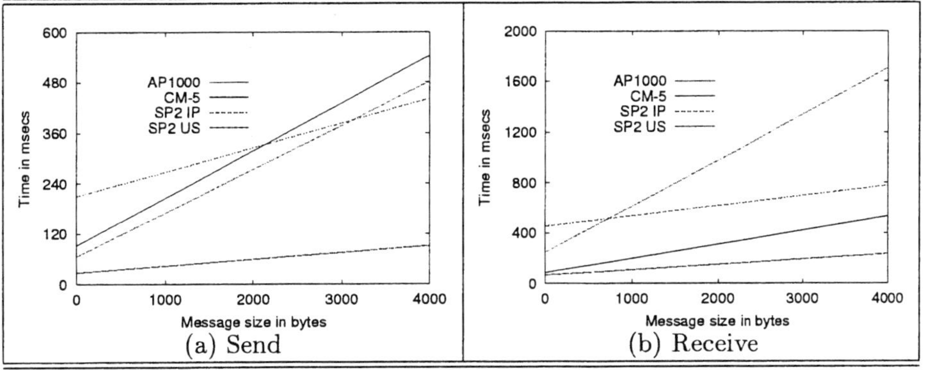

FIG. 3.4. *A comparison of send and receive overheads*

Processor Count (P)	Message Size (bytes)		
	2000	4000	8000
2	0.003562	0.006442	0.007849
4	0.004441	0.007545	0.011505
6	0.005856	0.009843	0.013747
8	0.008345	0.014372	0.029795
10	0.010349	0.016589	0.036853

(a) Communication overhead (seconds)

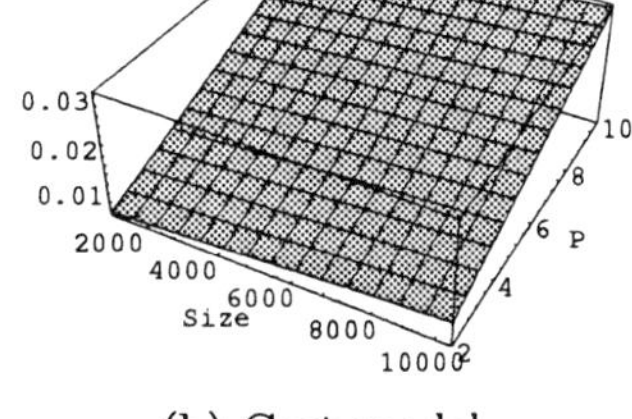

(b) Cost model

FIG. 3.5. *Send overhead for a pair of workstations*

Alpha workstations. To simulate the network contention, we vary the number of workstations used in our experiments when performing the Ping and Ping-Pang experiments. Some experimental results are given in Figure 3.5 (a) and illustrated in Figure 3.5 (b). From Figure 3.5 (b), we can see that the send overhead increases as both the granularity of messages and the number of workstations increase. This shows that network contention has a great effect on communication overhead for NOWs.

4. Two Communication Issues. In this section, we discuss two problems that arise in the context of reducing communication overheads for data-parallel programs.

4.1. First Send v.s. Subsequent Sends. In our experiments conducted on all four multicomputer systems, the overhead of the first send in a processor is always higher than that of a subsequent send. Table 4.1 lists the message communication times for a message of 1K bytes on the four systems. The first send overhead is about four times as expensive as a subsequent send and the overheads of all subsequent calls do not fluctuate significantly. The higher first send overhead appears to be caused by operations such as protocol processing and buffer management.

Send	SP2 (US)	CM-5	AP1000	NOWs
1st	0.000151	0.001107	0.000837	0.009526
2nd	0.000045	0.000285	0.000144	0.002380
3rd	0.000043	0.000261	0.000176	0.002421
4th	0.000051	0.000259	0.000130	0.002830
5th	0.000040	0.000302	0.000151	0.002572

TABLE 4.1
Send overhead (secs): first send v.s. subsequent sends

To investigate the impact of the first send overhead on the performance of a data-parallel program, we partition a 2D SOR program by loop tiling and execute the tiles in the SPMD paradigm. The SPMD code has the form given in Figure 4.1.

```
for each tile T allocated to processor P_k do
    if P_k ≠ 1 do
        receive the data that T depends from P_{k-1}
    compute(T)
    if P_k ≠ P do
        send the data dependent on T to P_{k+1}
```

FIG. 4.1. *Skeleton SPMD for the tiled 2D-SOR program*

Each processor executes the SPMD code on different data sets. After completing its first tile, processor P_1 sends P_2 the data that P_2 depends. Then P_1 and P_2 can run in parallel. The remaining processors work in a similar fashion. The execution of the SPMD code is illustrated in Figure 4.2 (a). In each processor, the first send incurs a larger overhead than a subsequent send. This has degraded the overall performance. Furthermore, the situation gets worse as the number of processors used increases.

A simple technique, called *PreLink*, is effective in reducing the impact of the first send overhead. When the first processor starts to compute its tiles, each of the remaining processors sends a small message to its right neighbouring processor. By overlapping all expensive first sends, the overall performance can be improved. Figure 4.2 (b) illustrates the execution of the program with the PreLink technique.

We ran the 2D SOR code using the data set 1000×1000 on CM-5 and AP1000 with and without *PreLink* (SP2 and NOWs were not used due to their lack of enough processors). As Table 4.2 shows, *PreLink* achieves a 17.88% performance improvement. Better performance improvements are expected when more processors are used.

4.2. Blocking Send v.s. Non-blocking Send. A blocking send will not complete until either a matching receive has been posted or the data has been safely copied out of the sender buffer. A non-blocking send, on the other hand,

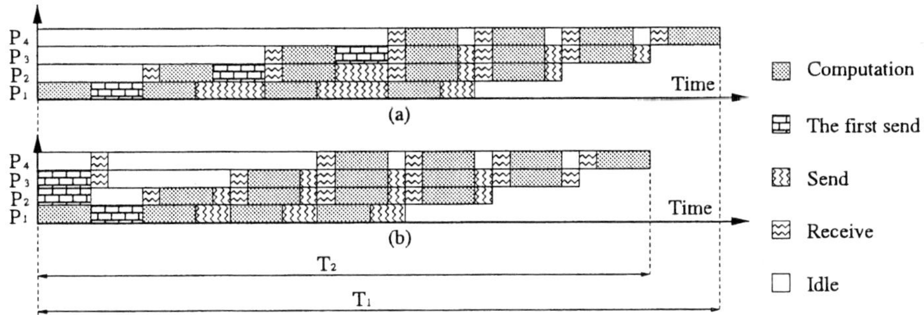

FIG. 4.2. *Execution of 2D SOR: Non-PreLink v.s. PreLink*

System	Processor(P)	Non-PreLink	PreLink	Improvement
CM-5	4	0.105218	0.104844	0.36%
	8	0.097440	0.091276	6.33%
	16	0.088412	0.080320	9.15%
AP1000	8	0.985820	0.983941	0.19%
	16	0.556008	0.534082	3.94%
	64	0.332518	0.273065	17.88%

TABLE 4.2
Performance of 2D SOR (secs): Non-PreLink v.s. PreLink

may proceed concurrently with computations done at the sender after the send was initiated and before it completed. Therefore, nonblocking communication can potentially improve performance on many systems by overlapping communication and computation.

In practice, nonblocking communication can degrade the performance of some programs in a specific architecture. We coded two versions of the 2D SOR program: one using blocking send calls and the other using nonblocking send calls. We executed both programs on SP2 and CM-5 (CAP-II on AP1000 and PVM on our NOWs do not support nonblocking send calls). The results are given in Table 4.3.

In our experiments, the performance of the nonblocking program is consistently worse. On the average, the performance degradation is 10%. This is mainly due to the lack of a dedicated communication controller in both machines. Unlike the Intel Paragon 250 in which each node has a 50MHz *i*860 main processor and a separate *i*860 co-processor specializing in handling communication [16], the nodes on our systems do not have such a co-processor to support communication and computation overlap. All communication overheads have to be charged to the main processor of each node.

Therefore, care must be taken when nonblocking communication is used. In

System	Processor (P)	Blocking	Non-blocking	Degradation
SP2(US)	4	0.226571	0.229782	-1.41%
	8	0.127212	0.138170	-8.62%
	10	0.120490	0.122172	-1.39%
CM-5	8	0.097448	0.098562	-11.43%
	16	0.088412	0.099250	-12.25%
	32	0.088101	0.103352	-17.27%

TABLE 4.3
Performance (seconds) of 2D SOR: Blocking v.s. Non-blocking

the case of data-parallel programs, blocking communication may be preferred. Non-blocking communication can enhance performance when an intelligent communication controller is available in each processor to handle communication.

5. Related Work. Communication overhead is an important issue in distributed computing. Both programmers and parallelizing compilers need a cost model to estimate communication costs in order to achieve good performance.

The well-known *LogP* model is presented in [5]. Later, two extensions, *LoGP* [17] and *LoPC* [7], are proposed. While *LogP* provides an elegant abstraction to model various communication modes, neither the sending processor nor the receiving processor can use the model directly. In addition, the *LogP* parameters, L, o and g, are nontrivial to measure in a real machine. Finally, neither *LogP* nor *LoGP* explicitly uses the fourth parameter, the number of processors (P), to model network contention. The *LoPC* model approximates the cost of network contention using P. However, it is different from our cost model in the following aspects:

- While our cost model considers the contention on the network as shared by all processors, the *LoPC* model deals with the contention on the message processing resources on individual processors.
- *LoPC* is more accurate than ours at the expense of more algorithm parameters in the model and can be algorithm-dependent and difficult to use.
- *LoPC* inherits the difficulty in measuring its parameters from *LogP* .

Network contention is also considered in the *BSP* (*Bulk Synchronous Parallel*) model [20]. However, *BSP* represents network contention only in the form of tables rather than closed-form expressions. Therefore, *BSP* cannot be used in the case when the optimal number of processors for applications is expected to be derived [4].

The other related work is given as follows. Many software systems have been developed to support the message-passing paradigm, such as PVM [9], MPI [21] and COBRA (Common Object Broker Request Architecture) [18]. Fatoohi evaluated the performance of the communication software systems on a network of SGI workstations connected by four networks: Ethernet, FDDI, HiPPI and ATM [6].

6. Conclusion. In this paper, we have investigated the problem of estimating communication overhead on distributed memory machines. We started by discussing how the *LogP* model can be used to estimate communication overhead on distributed memory machines. To overcome some of its limitations, a communication cost model is proposed. In the proposed cost model, the effects of send, receive and network contention on overall communication overhead are quantified separately. Such a cost model is expected to be useful for multicomputers as well as networks of workstations. We presented Ping and Ping-Pang algorithms used for determining the communication parameters in our cost model in a given machine. We have applied these algorithms to build the concrete cost models for four commercial multicomputer systems, IBM SP2, TMC CM-5, FUJITSU AP1000 and a network of DEC Alpha workstations. By quantifying the communication overheads at application level, our cost model can be used to estimate communication overhead for various communication modes.

We discussed two communication issues that may affect the performance of parallel programs (especially SPMD codes). First, we observed that on many distributed systems, the communication overhead for the first send is much higher than that of a subsequent send. A simple technique is proposed to reduce the effect of the first send overhead on overall program performance. This technique is validated using a 2D SOR program. Second, a comparison of blocking and nonblocking send calls was made using the SOR program. Blocking communication is preferred if the machine used does not have any dedicated hardware support for enhancing communication performance.

Acknowledgements. The authors wish to thank QPSF, SACPC and ANUSF for allowing us to access their multicomputer systems. This work is supported by an Australian Research Council Grant A49600987.

REFERENCES

[1] R. Andonov and S. Rajopadhye, *Optimal tiling of two-dimensional uniform recurrences*, Tech. Rep. 97–01, LIMAV, Université de Valenciennes, Jan. 1997.

[2] S. Bokhari, *Communication overhead on the Inter Paragon, IBM-SP2 and meiko CS-2*, Tech. Rep. TR-28, NASA Langley Research Center, August 1995.

[3] J. M. Chambers and T. M. Hastie, *Statistical Models in Splus*, 0-534-16764-0, Wadsworth Inc., 1992.

[4] S. Chen and J. Xue, *Issues of tiling double loops on distributed memory machines*, in the 5th Annual Australasian Conference on Parallel And Real-Time Systems (PART'98), Adelaide, Australia, Sept. 1998, pp. 377–388.

[5] D. E. Culler, et al, *LogP: Towards a realistic model of parallel computation*, in The 4th ACM SIGPLAN Symposium on Principles and Practice of Parallel Programming, 1993, pp. 262–273.

[6] R. Fatoohi, *Performance evaluation of communication software systems for distributed computing*, Tech. Rep. NAS-96-006, NASA Ames Research Center, June 1996.

[7] M. Frank, M. Vemon, and A. Agarwal, *LoPC: Modelling contention in parallel algorithms*, in The 4th ACM SIGPLAN Symposium on Principles and Practice of Parallel Programming (PPoPP'97), Las Vegas, 1997.

[8] Fujitsu, *AP1000 Users' Guide*, 1994.

[9] A. Geist, et al, *PVM – Parallel Virtual Machine: A Users' Guide and Tutorial for Networked Parallel Computing*, 0-262-57108-0, The MIT Press, 1994.

[10] M. Gupta and P. Banerjee, *Compile-time estimation of communication cost on multcomputers*, in Proceedings of the 6th International Parallel Processing Symposium, Beverly Hills, CA, March 1992, pp. 470–475.

[11] K. Hwang, *Advanced Computer Architecture: Parallelism, Scalability and Programmability*, MIT Press and McGraw-Hill Inc., 1993.

[12] IBM, *IBM Parallel Environment for AIX (V2.1.0)*, 1995.

[13] N. Nupairoj and L. Ni, *Performance evaluation of some mpi implementations on workstation clusters*, in the 1994 Scalable Parallel Libraries Conference (SPLC94), Oct. 1994, pp. 98–105.

[14] ———, *Performance metrics and measurement techniques of collective communication services,*, in The 1st International Workshop on Communication and Architectural Support for Network-Based Parallel Computing (CANPC'97), San Antonio, TX, Feb. 1997, pp. 212–226.

[15] H. Ohta, Y. Saito, M. Kainaga, and H. Ono, *Optimal tile size adjustment in compiling for general DOACROSS loop nests*, in Supercomputing '95, ACM Press, 1995, pp. 270–279.

[16] D. J. Palermo, *Compiler Techiques For Optimising Communication and Data Distribution for Distributed Multicomputers*, PhD thesis, Dept. Computer Science, Nov. 1996.

[17] E. A. R. Martin, *Effects of communication latency, overhead, and bandwidth in a cluster architecture*, in International Symposium on Computer Architecture, Denver, CO, June 1997.

[18] D. C. Schmidt and T. Suda, *A high-performance endsystem architecture for real-time CORBA*, IEEE Communications Magazine, 14 (1997).

[19] Thinking Machines Cooperation, *CMMD User's Guide (V3.3)*, 1994.

[20] L. Valiant, *A bridging model for parallel computation*, Communications of the ACM, 33 (1990), pp. 103–111.

[21] E. L. W. Gropp and A. Skjellum, *Using MPI: Portable Parallel Programming with the Message-Passing Interface*, 0-262-57104-8, The MIT Press, 1994.

RANK ORDER FILTERING ON MASSIVELY-PARALLEL SINGLE-BIT MESH PROCESSOR ARRAYS

HONGCHI SHI* AND HONGZHENG LI†

Abstract. Rank order filters have a wide variety of applications in image processing as an effective tool for removing noise in images without severely distorting abrupt changes in the images. The k-th rank filter with an $m \times m$ window sets each pixel the k-th smallest value of the m^2 pixels in its $m \times m$ neighborhood. The median filter is the most commonly used special case of rank order filters. Rank order filtering requires intensive computation. In this paper, we consider implementation of rank order filters on massively-parallel single-bit mesh-connected computers such as the Lockheed Martin Parallel Algebraic Logic (PAL) computer.

We design rank filtering algorithms using the threshold decomposition and radix splitting techniques. We map an $n \times n$ image onto an $n \times n$ single-bit mesh array with one pixel per processing element (PE). Suppose that the maximum pixel value is L. Using the threshold decomposition technique, we can implement any rank filter in $O((m \log m + \log L)L)$ time on a mesh processor array with $O(\log L + \log m)$ bits of space on each PE. Using the radix splitting technique, we can implement any rank order filter in $O(m^2 \log m \log L)$ time on a mesh processor array with $O(\log L + m^2)$ bits of space on each PE. If the local space on each PE is limited to $O(\log L + \log m)$ bits, we can have an $O(m^2(\log L + \log m) \log L)$ implementation. We also present some experimental results of the implementation of those algorithms on the PAL computer.

Key words. rank order filter, threshold decomposition, radix splitting, parallel algorithm, mesh processor array

1. Introduction.

Digital filtering has many applications in a wide variety of fields such as biomedical engineering, radar, speech and data communication, and image processing. Linear filters have been the main tool for signal and image processing for a long time. However, linear filters tend to distort abrupt changes in non-stationary signals such as the edge regions in images and can not remove impulsive noise totally. In those cases, nonlinear filters such as rank order filters are more appropriate.

A k-th rank filter with an $m \times m$ window sets each pixel the k-th smallest value of the m^2 pixels in its $m \times m$ neighborhood. Many researchers have studied the most commonly used special rank filter (i.e., the median filter) and developed fast algorithms for it on serial and parallel computers [1, 2, 3, 4, 5, 6, 7, 8]. Some researchers has also studied parallel implementation of rank order filters [9, 10, 11, 12]. However, most of them consider implementations using VLSI and FPGA. In this paper, we are developing algorithms for rank order filters on a general mesh-connected processor array and evaluating their performance.

Rank filters involve intensive computation. Suppose the source image is of size $n \times n$ and its maximum gray level is L. The k-th rank filter has window size $m \times m$. The traditional method sorts all the m^2 pixels in the window and

*Department of Computer Engineering & Computer Science, University of Missouri-Columbia, Columbia, MO 65211 (shih@missouri.edu).

†VM Labs, Inc., 520 San Antonio Road Mountain View, CA 94040 (hli@vmlabs.com).

picks the k-th smallest value for each pixel, thus needing $O(n^2m^2 \log m \log L)$ bit operations on a serial computer, since sorting m^2 $(\log L)$-bit values needs $O(m^2 \log m^2 \log L)$ bit operations. Various fast algorithms have been proposed for the median filter [1, 2, 3, 4, 5]. The best serial algorithm combines the moving window technique and a balanced search tree data structure, reducing the time to $O(n^2m \log m \log L)$ bit operations. Parallel algorithms for the median filter on mesh-connected computers have been developed to further reduce the time complexity [6]. The SIMD mesh architecture is considered as a natural parallel architecture for image processing, since it directly mirrors image data structures and can be efficiently implemented in hardware. Most of image processing computers based on the SIMD mesh architecture are massively-parallel single-bit processor arrays [13, 14, 15, 16, 17, 18, 19, 20, 21].

In this paper, we design parallel algorithms for the general case of rank order filters on massively-parallel single-bit mesh processor arrays such as the Lockheed Martin Parallel Algebraic Logic (PAL) computer. We map an $n \times n$ image onto an $n \times n$ mesh array with one pixel per processing element (PE). Using the threshold decomposition technique, we design an $O((m \log m + \log L)L)$ time algorithm for any rank filter of size $m \times m$ on a mesh processor array with $O(\log L + \log m)$ bits of local space on each PE. Using the radix splitting technique, we design an $O(m^2 \log m \log L)$ time algorithm for any rank order filter on a mesh processor array with $O(\log L + m^2)$ bits of space on each PE. If the local space on each PE is limited to $O(\log L + \log m)$ bits, we have an $O(m^2(\log L + \log m) \log L)$ time algorithm. We also present some experimental results of the implementation of those algorithms on the PAL computer.

2. Massively-Parallel Single-Bit Mesh Processor Arrays. An $n \times n$ single-bit massively-parallel mesh processor array computer is composed of n^2 processing elements (PEs) connected through an $n \times n$ mesh interconnection network as shown in Figure 2.1. An example of such an SIMD mesh processor array is the PAL computer [20, 21]. We denote the PE in the i-th row and j-th column of the mesh by PE(i, j) with PE(0,0) in the top left corner of the mesh.

The PEs are single-bit and operate in a single instruction stream multiple data stream (SIMD) mode, with all control signals coming from a single control unit. The control unit can broadcast a 1-bit constant to all the PEs in a unit time. Each PE is connected to its four neighbors. The links in the interconnection network are also single-bit and unidirectional. In a unit time, all the PEs may transmit a bit of data to their directly connected PEs in a specified direction. Thus, it takes $O(l)$ time to perform an arithmetic or logic operation involving l-bit operands. It also takes $O(l)$ time to move an l-bit value from each PE to its neighboring PE in a specified direction.

For an $n \times n$ image $\mathbf{a}$ on an $n \times n$ mesh processor array, each pixel $\mathbf{a}(x, y)$ is mapped to PE(x, y). Neighboring pixels are mapped to neighboring PEs. For images larger than the processor array, we can process the images block by block with the same efficiency.

For a neighborhood operation, we move all the pixels in the neighborhood

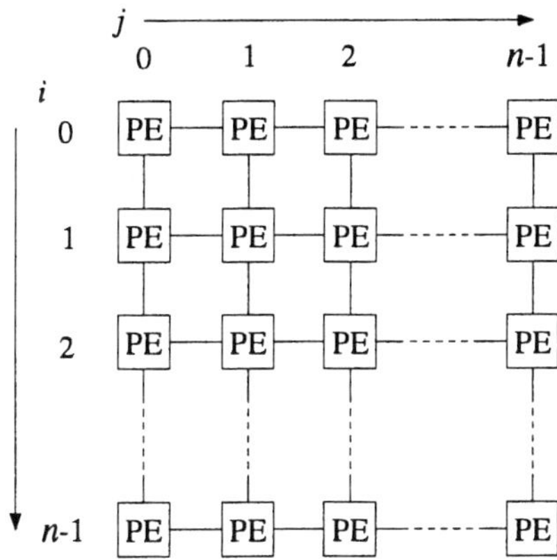

FIG. 2.1. *A mesh processor array*

around each pixel $\mathbf{a}(x, y)$ to PE(x, y) and perform some operation on PE(x, y) once a pixel is moved in. In order to accomplish the moving process in the SIMD mode, we design moving paths for the neighborhood to move the pixels in the neighborhood of $\mathbf{a}(x, y)$ one at a time to PE(x, y) along some path. For a row neighborhood, we use two moving paths as illustrated in Figure 2.2(a). The arrows in the figure denote the moving directions. Similarly, we use two moving paths for a column neighborhood as shown in Figure 2.2(b). For a square $m \times m$ neighborhood, we use a single moving path starting at the center pixel and including all the pixels in the neighborhood in a spiral clockwise order. For example, the moving path for the 3×3 neighborhood is $(0,0)$, $(0,1)$, $(1,1)$, $(1,0)$, $(1,-1)$, $(0,-1)$, $(-1,-1)$, $(-1,0)$, $(-1,1)$ as shown in Figure 2.2(c). To obtain moving paths for more complicated neighborhoods, we can use the similar techniques used for finding convolution paths for those neighborhoods [22, 23].

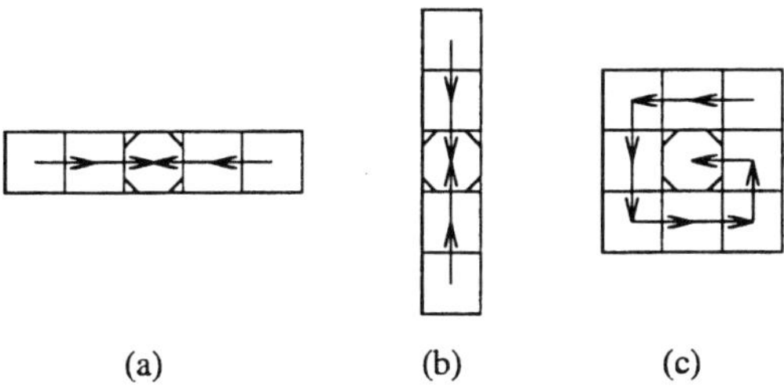

FIG. 2.2. *Illustration of moving paths for different neighborhoods*

It is easy to figure out the shift operation for each moving step for a row or column neighborhood. For the $m \times m$ neighborhood, we can use the following procedure to determine the shift operation for each moving step. The pixels in the neighborhood are ordered according to their positions in the moving path.

```
for (p = 1, g = 1; p < m^2; g++)
    if (g%2) {
        for (l = 0; l < g && p < m^2; l++, p++)
            Shift left to get the p-th pixel;
        for (l = 0; l < g && p < m^2; l++, p++)
            Shift up to get the p-th pixel;
    } else {
        for (l = 0; l < g && p < m^2; l++, p++)
            Shift right to get the p-th pixel;
        for (l = 0; l < g && p < m^2; l++, p++)
            Shift down to get the p-th pixel;
    }
```

The entire moving process takes $O(m^2)$ time for moving a bit of each pixel in the $m \times m$ neighborhood.

3. Rank Filtering by Threshold Decomposition. Median Filtering can be achieved by threshold decomposition [24, 25]. Median filtering a gray level image is equivalent to decomposing the original image into binary images, filtering each binary image with a binary median filter, and then reversing the decomposition by adding up the binary median filtering results. The threshold decomposition technique decomposes the complicated gray-level median filtering problem to a set of simple binary median filtering problems.

Here, we use the threshold decomposition technique to implement rank order filters on single-bit mesh processor arrays. Suppose that the source image $\mathbf{a}$ is on an $n \times n$ point set, the maximum pixel value in the image is $L-1$, the window size of the rank filter is $m \times m$. Let $\mathbf{b}$ be the k-th rank filtering result of $\mathbf{a}$. That is, $\mathbf{b}(x, y)$ is the k-th smallest value of the m^2 pixels in the $m \times m$ neighborhood of $\mathbf{a}(x, y)$. Define the level l threshold decomposition of the pixel $\mathbf{a}(x, y)$ to be

$$\mathbf{a}_l(x, y) = \begin{cases} 1 & \text{if } \mathbf{a}(x, y) \geq l \\ 0 & \text{if } \mathbf{a}(x, y) < l, \end{cases}$$

where $0 \leq x, y \leq n-1$ and $1 \leq l \leq L-1$.

Applying the k-th rank filter to each thresholded binary image $\mathbf{a}_l$, we obtain a binary image $\mathbf{b}_l$ as follows:

$$\begin{aligned} \mathbf{b}_l(x, y) &= k\text{-th smallest value of } \{\mathbf{a}_l(x+i, y+j) : -\lfloor \frac{m-1}{2} \rfloor \leq i, j \leq \lfloor \frac{m}{2} \rfloor\} \\ &= \begin{cases} 1 & \text{if } \sum_{-\lfloor \frac{m-1}{2} \rfloor \leq i,j \leq \lfloor \frac{m}{2} \rfloor} \mathbf{a}_l(x+i, y+j) > m^2 - k \\ 0 & \text{if } \sum_{-\lfloor \frac{m-1}{2} \rfloor \leq i,j \leq \lfloor \frac{m}{2} \rfloor} \mathbf{a}_l(x+i, y+j) \leq m^2 - k. \end{cases} \end{aligned}$$

We can see that $\mathbf{b}_l(x, y) = 1$ if and only if there are fewer than k pixels of $\mathbf{a}$ in the $m \times m$ neighborhood of $\mathbf{a}(x, y)$ having values smaller than l. Furthermore, $\mathbf{b}_l(x, y) = 1$ implies $\mathbf{b}_{l'}(x, y) = 1$ for all $1 \leq l' \leq l$. Thus, $\max\{l : \mathbf{b}_l(x, y) = 1\} = \sum_{l=1}^{L-1} \mathbf{b}_l(x, y)$.

The k-th rank filtering result $\mathbf{b}$ of $\mathbf{a}$ can be obtained from $\mathbf{b}_l$ as follows:

$$\begin{aligned}\mathbf{b}(x,y) &= k\text{-th smallest value of } \{\mathbf{a}(x+i,y+j) : -\lfloor\frac{m-1}{2}\rfloor \le i,j \le \lfloor\frac{m}{2}\rfloor\} \\ &= \max\{l : \text{Fewer than } k \text{ pixels in } \mathbf{a}(x,y)\text{'s } m\times m \text{ neighborhood} \\ &\qquad\qquad \text{have values smaller than } l\} \\ &= \max\{l : \mathbf{b}_l(x,y) = 1\} \\ &= \sum_{l=1}^{L-1} \mathbf{b}_l(x,y).\end{aligned}$$

We observe that most of the operations involved in the k-th rank filtering using threshold decomposition are neighborhood operations on binary images. Furthermore, they are applied to every pixel of the source image. Thus, we can effectively use the threshold decomposition technique to compute any rank filter on SIMD single-bit mesh processor arrays. We simply decompose the source image into $L-1$ binary images; perform binary rank filtering on the decomposed binary images by counting and thresholding; and accumulate the binary rank filtered result to the final result for each pixel.

Obviously, it takes $O(\log L)$ time to compute each $\mathbf{a}_l$ on a single-bit mesh processor array. Thus, the decomposition process takes $O(L\log L)$ time in total.

To compute $\mathbf{b}_l$ from $\mathbf{a}_l$, we first count the 1's in the $m\times m$ neighborhood of each $\mathbf{a}_l(x,y)$. Since $\sum_{-\lfloor\frac{m-1}{2}\rfloor\le i,j\le\lfloor\frac{m}{2}\rfloor} \mathbf{a}_l(x+i,y+j) = \sum_{i=-\lfloor\frac{m-1}{2}\rfloor}^{\lfloor\frac{m}{2}\rfloor}(\sum_{j=-\lfloor\frac{m}{2}\rfloor}^{\lfloor\frac{m}{2}\rfloor} \mathbf{a}_l(x+i,y+j)) = \sum_{i=-\lfloor\frac{m}{2}\rfloor}^{\lfloor\frac{m}{2}\rfloor} \mathbf{a}'_l(x+i,y)$, where $\mathbf{a}'_l(x,y) = \sum_{j=-\lfloor\frac{m-1}{2}\rfloor}^{\lfloor\frac{m}{2}\rfloor} \mathbf{a}_l(x,y+j)$, we can decompose the counting process by counting in columns and then in rows. Thus, the counting process needs $O(m\log m)$ bit operations, since the sum is at most m^2. We then perform a thresholding operation to compute $\mathbf{b}_l$ from the counts of 1's, which can be done in $O(\log m)$ time. In summary, computing each $\mathbf{b}_l$ takes $O(m\log m)$ time on single-bit mesh processor arrays.

Adding $\mathbf{b}_l$ $(1\le l\le L-1)$ up pixel-wise, we obtain $\mathbf{b}$. This process can be done in $O(L\log L)$ time.

Thus, the algorithm using threshold decomposition takes $O((m\log m + \log L)L)$ time on a mesh processor array with $O(\log L + \log m)$ bits of local space on each PE.

4. Rank Filtering by Radix Splitting. Another basic technique for fast implementation of the median filter is radix splitting [2]. The basic idea is represent the pixels in radix-2 format, split the pixels in the neighborhood of each pixel into sets according to their values in each bit repeatedly, and determine the bits of the median according to the sizes of those sets.

Here, we apply the radix splitting technique to implement the k-th rank order filter on single-bit mesh processor arrays. Let $\mathbf{b}$ be the k-th rank filtering result of $\mathbf{a}$. Suppose the binary representation of each pixel $\mathbf{a}(x,y)$ is $\mathbf{a}^{\log L-1}(x,y)\ \mathbf{a}^{\log L-2}(x,y)\cdots\mathbf{a}^0(x,y)$ and the k-th smallest pixel in $\mathbf{a}(x,y)$'s $m\times m$ neighborhood is represented as $\mathbf{b}^{\log L-1}(x,y)\mathbf{b}^{\log L-2}(x,y)\cdots\mathbf{b}^0(x,y)$.

Let $G_{b^{\log L-1}b^{\log L-2}\dots b^l}(x,y)$ be the set of the pixels in the $m\times m$ neighborhood of $\mathbf{a}(x,y)$ with the first $\log L - l$ most significant bits equal to $b^{\log L-1}b^{\log L-2}\dots b^l$. That is,

$$\begin{aligned}&G_{b^{\log L-1}b^{\log L-2}\dots b^l}(x,y)\\&=\{\mathbf{a}(x+i,y+j) : \mathbf{a}^{\log L-1}(x+i,y+j)\cdots\mathbf{a}^l(x+i,y+j) = b^{\log L-1}\dots b^l,\\&\quad -\lfloor\frac{m-1}{2}\rfloor \le i,j \le \lfloor\frac{m}{2}\rfloor\}.\end{aligned}$$

We first split the m^2 pixels in the neighborhood of $\mathbf{a}(x,y)$ into two sets $G_0(x,y)$ and $G_1(x,y)$. Obviously, all the numbers in $G_1(x,y)$ are greater than those in $G_0(x,y)$. If $|G_0(x,y)| \ge k$, we know that $\mathbf{b}(x,y)$ must be in $G_0(x,y)$, thus $\mathbf{b}^{\log L-1}(x,y) = 0$; otherwise, $\mathbf{b}(x,y)$ must be in $G_1(x,y)$, thus $\mathbf{b}^{\log L-1}(x,y)=1$. If $\mathbf{b}^{\log L-1}(x,y) = 0$, we can further split $G_0(x,y)$ into $G_{00}(x,y)$ and $G_{01}(x,y)$ to determine $\mathbf{b}^{\log L-2}(x,y)$ by checking whether the k-th element of $G_0(x,y)$ is in $G_{00}(x,y)$ or $G_{01}(x,y)$. If $\mathbf{b}^{\log L-1}(x,y) = 1$, we can further split $G_1(x,y)$ into $G_{10}(x,y)$ and $G_{11}(x,y)$ to determine $\mathbf{b}^{\log L-2}(x,y)$ by checking whether the $(k-|G_0(x,y)|)$-th element of $G_1(x,y)$ is in $G_{10}(x,y)$ or $G_{11}(x,y)$. Repeating the radix splitting process $\log L$ times, we can determine all the bits of $\mathbf{b}(x,y)$.

We observe that one set is needed at any time, i.e., $G_{\mathbf{b}^{\log L-1}(x,y)\cdots\mathbf{b}^{l+1}}$ $(x,y)(x,y)$ is needed to determine $\mathbf{b}^l(x,y)$. Furthermore, the set $G_{\mathbf{b}^{\log L-1}(x,y)\cdots}$ $\mathbf{b}^{l+1}(x,y)\mathbf{b}^l(x,y)(x,y)$ can be computed either directly by definition or recursively by

$$\begin{aligned}&G_{\mathbf{b}^{\log L-1}(x,y)\cdots\mathbf{b}^{l+1}(x,y)\mathbf{b}^l(x,y)}(x,y)\\&=\{\mathbf{a}(x',y')\in G_{\mathbf{b}^{\log L-1}(x,y)\cdots\mathbf{b}^{l+1}(x,y)}(x,y) : \mathbf{a}^l(x',y') = \mathbf{b}^l(x,y)\}\end{aligned}$$

after $\mathbf{b}^l(x,y)$ is determined. An algorithm based on the radix splitting technique is as follows:

```
for (l = log L - 1; l ≥ 0; l - -) {
    Compute c = |G_{b^{log L-1}(x,y)...b^{l+1}(x,y)0}(x, y)|;
    if (c ≥ k)
      b^l(x, y) = 0;
    } else {
      b^l(x, y) = 1;
      k = k - c;
    }
    Compute G_{b^{log L-1}(x,y)...b^{l+1}(x,y)b^l(x,y)}(x, y);
}
```

The sets for different pixels may be different in size and elements. On an SIMD processor array, however, we have to use the same instruction in each step. To maintain the SIMD fashion, we use a mask to indicate which pixels in the neighborhood belong to the set under consideration and process the pixels in the neighborhood uniformly in each step. We can either store a mask or compute the mask on the fly. Thus, we design two radix splitting algorithms for rank filtering on single-bit mesh processor arrays. They have different requirements for local space on each PE.

4.1. First Radix Splitting Algorithm. In the first algorithm using radix splitting, we store a mask $\mathbf{m}(x,y)$ of m^2 bits on each PE for the $m \times m$ neighborhood. We order the pixels in the neighborhood of $\mathbf{a}(x,y)$ as $\mathbf{a}_0(x,y)$, $\mathbf{a}_1(x,y)$, $\cdots$, $\mathbf{a}_{m^2-1}(x,y)$ according to the moving path shown in Figure 2.2(c). The p-th bit $\mathbf{m}_p(x,y)$ indicates whether $\mathbf{a}_p(x,y)$, the p-th pixel in the neighborhood of $\mathbf{a}(x,y)$, belongs to the set under consideration.

During the filtering process, each PE updates its mask according to its current set. In determining $\mathbf{b}^l(x,y)$, $\mathbf{m}_p(x,y) = 1$ implies $\mathbf{a}_p(x,y) \in G_{\mathbf{b}^{\log L-1}(x,y)\cdots}$ $\mathbf{b}^{l+1}(x,y)(x,y)$. The mask $\mathbf{m}(x,y)$ is updated according to $G_{\mathbf{b}^{\log L-1}(x,y)\cdots\mathbf{b}^{l+1}}$ $(x,y)\mathbf{b}^l(x,y)(x,y)$ after $\mathbf{b}^l(x,y)$ is determined.

To compute $|G_{\mathbf{b}^{\log L-1}(x,y)\cdots\mathbf{b}^{l+1}(x,y)0}(x,y)|$, we count the number of 0-pixels in the neighborhood whose corresponding mask bits are 1. We use the moving procedure to move the l-th bit of each $\mathbf{a}_p(x,y)$ in the neighborhood of $\mathbf{a}(x,y)$ to PE(x,y) one by one, and increment the count by 1 if $\mathbf{m}_p(x,y) = 1$ and $\mathbf{a}_p(x,y) = 0$. This process takes $O(m^2 \log m)$ time, since there are m^2 pixels in the neighborhood and the count is at most $\log m^2$ bits long. We then compare the count with k to determine $\mathbf{b}^l(x,y)$, which takes $O(\log m)$ time.

To update the mask after the $\mathbf{b}^l(x,y)$ is determined, we move the l-th bit of each $\mathbf{a}_p(x,y)$ in the neighborhood to PE(x,y) again. Once the bit is moved in, we reset $\mathbf{m}_p(x,y)$ if $\mathbf{m}_p(x,y) = 1$ and $\mathbf{a}_p^l(x,y) \neq \mathbf{b}^l(x,y)$. This process takes $O(m^2)$ time.

The first radix splitting algorithm is described as follows:

1. Initialize $\mathbf{m}_p(x,y)$ $(p = 0, 1, \cdots, m^2-1)$ as 1 on each PE(x,y);
2. For $l = \log L - 1, \log L - 2, \cdots, 0$,
   2.1 Move the l-th bit of each $\mathbf{a}_p(x,y)$ in $\mathbf{a}(x,y)$'s $m \times m$ neighborhood to PE(x,y) and increment the count by 1 if $\mathbf{m}_p(x,y) = 1$ and$\mathbf{a}_p^l(x,y) = 0$;
   2.2 If the count is no smaller than k
      Set $\mathbf{b}^l(x,y) = 0$ on each PE(x,y);
   else
      Set $\mathbf{b}^l(x,y) = 1$ on each PE(x,y);
      Subtract the count from k;
   2.3 Move the l-th bit of each $\mathbf{a}_p(x,y)$ in the neighborhood to PE(x,y) and reset $\mathbf{m}_p(x,y)$ if $\mathbf{m}_p(x,y) = 1$ and $\mathbf{a}_p^l(x,y) = \mathbf{b}^l(x,y)$;

It is easy to see that the above algorithm requires $O(\log L + m^2)$ bits of local space on each PE and takes $O(m^2 \log m \log L)$ time on single-bit mesh processor arrays.

4.2. Second Radix Splitting Algorithm. The first radix splitting algorithm has a rather large local space requirement for each PE. However, most of massively-parallel single-bit processor arrays are designed with limited local memory space. For example, the PAL computer has only 192 bits of local memory on each PE. The first radix splitting algorithm is not practical on such a single-bit mesh processor array when the window size is large.

To reduce the local space requirement, we design another algorithm that computes the mask on the fly in stead of storing it. To compute $|G_{\mathbf{b}^{\log L-1}(x,y)\cdots}$ $\mathbf{b}^{l+1}(x,y)0(x,y)|$ for $\mathbf{b}^l(x,y)$, we move the first $\log L - l$ most significant bits $\mathbf{a}_p^{\log L-1}(x,y)\cdots\mathbf{a}_p^l(x,y)$ of each $\mathbf{a}_p(x,y)$ in $\mathbf{a}(x,y)$'s neighborhood of to PE(x,y). With $\mathbf{a}_p^{\log L-1}(x,y)\cdots\mathbf{a}_p^l(x,y)$ moved in, if $\mathbf{a}_p^{\log L-1}(x,y)\cdots\mathbf{a}_p^l(x,y)$ $= \mathbf{b}^{\log L-1}(x,y)\cdots\mathbf{b}^{l+1}(x,y)0$, we add 1 to the count. Thus, for each l, we need to perform $O(m^2((\log L - l)\log m))$ bit operations. The second radix splitting algorithm is described as follows:

For $l = \log L - 1, \log L - 2, \cdots, 0$,
1. Move the first $\log L - l$ most significant bits $\mathbf{a}_p^{\log L-1}(x,y)\cdots\mathbf{a}_p^l(x,y)$ of each $\mathbf{a}_p(x,y)$ in the $m\times m$ neighborhood of $\mathbf{a}(x,y)$ to PE(x,y) and add 1 to the count if $\mathbf{a}_p^{\log L-1}(x,y)\cdots\mathbf{a}_p^l(x,y) = \mathbf{b}^{\log L-1}(x,y)\cdots\mathbf{b}^{l+1}(x,y)0$;
2. If the count is no smaller than k
   Set $\mathbf{b}^l(x,y) = 0$ on each PE(x,y);
   else
   Set $\mathbf{b}^l(x,y) = 1$ on each PE(x,y);
   Subtract the count from k;

Apparently, the second radix splitting algorithm only requires $O(\log m)$ bits of local space for the count on each PE in addition to the space for the filtering result. The total space requirement is $O(\log L + \log m)$. However, the algorithm increases the time complexity a little. Since it takes $O(m^2(\log L - l) + m^2\log m)$ time for each l, $l = \log L - 1, \cdots, 0$, the time complexity of the second algorithm is $O(m^2(\log L + \log m)\log L)$.

5. Experimental Results. To further evaluate the performance of the algorithms we have presented for rank order filters on single-bit mesh processor arrays, we have implemented them on the PAL computer.

The PAL computer is a massively parallel single-bit mesh-connected computer designed to perform real-time image processing [20, 21]. In order to allow a large number of PEs per chip, the processing elements are simple and operates at 40MHz. Each PE consists of single-bit registers and very limited local memory. Due to the limited local space on each PE, the first radix splitting algorithm can only work for rank filters of window size no larger than 3×3. Thus, we only compare the performance of the threshold decomposition algorithm with that of the second radix splitting algorithm. The first radix splitting algorithm should be a little faster than the second one, provided its required local space on each PE is available.

The images in the experiment have the same size as the PAL processor array. Thus, the image size is not a parameter in the performance of the algorithm. Since each of the algorithm involves exactly one image input to the processor array and one image output from the processor array, we exclude the image input and output time in all t-eh time complexity curves. Due to the fact that most images have 256 gray levels, we consider L as a constant in the performance comparison. In this case, the time complexity of the threshold decomposition algorithm is reduced to $T_t(m) = O(m\log m)$, while the time complexity of the second radix splitting algorithm is reduced to $T_r(m) = O(m^2\log m)$ (more accurately, $O(m^2 \log m + 8m^2)$). However, the multiplicative constant for $T_r(m)$ is 32 ($= L/\log L = 32$) times smaller than that for $T_t(m)$. The additive constant for $T_r(m)$ is also smaller than $T_t(m)$. Figure 5.1 shows the time in seconds of the two algorithm for m from 3 to 20. Also plotted in the figure are the ideal curves for $m \log m$ and $m^2 \log m$. As we can see, $T_t(m)$ matches $\frac{1}{128} m \log m + 0.32$, while $T_r(m)$ matches $\frac{1}{4096} m^2 \log m + \frac{1}{512} m^2 + 0.09$, which verifies our theoretical analysis.

To compare the performance of the two algorithms, we plotted their time complexity curves together in Figure 5.2. The time complexity $T_t(m)$ starts with a longer time than $T_r(m)$, but it grows slower than $T_r(m)$. When m is greater than 14, the threshold decomposition algorithm is faster than the second radix splitting algorithm. On the PAL computer, we should use the threshold decomposition algorithm for filters of window size larger than 14×14. For rank filters of smaller window size, we should use the second radix decomposition algorithm. For rank filters of very small window size such as 3×3, it may be better to use the first radix splitting algorithm.

6. Conclusion. We have designed three algorithms for rank order filters on single-bit mesh processor arrays and compared their performance on the PAL computer. For filtering an $n \times n$ image with maximum gray level L on an $n \times n$ single-bit mesh processor array with a rank filter of window size $m \times m$, the threshold decomposition algorithm takes $O((m \log m + \log L)L)$ time and requires only $O(\log L + \log m)$ bits of local space on each PE. The first algorithm using radix splitting takes $O(m^2 \log m \log L)$ time and requires $O(\log L + m^2)$ bits of local space on each PE, while the second algorithm takes $O(m^2(\log L + \log m) \log L)$ time and requires only $O(\log L + \log m)$ bits of local space on each PE. On a single-bit mesh processor array with limited local memory on each PE, the second radix splitting algorithm is more practical than the first one. When the filter window is large, the threshold decomposition algorithm is better than the radix splitting algorithms.

Acknowledgments. The research work presented in this paper was in part supported by the University of Missouri Research Board under grants RB-96-024 and RB-98-068. The authors also wish to thank the anonymous reviewers of this paper for their insightful comments and suggestions.

REFERENCES

[1] T. S. Huang, G. T. Yang, and G. Y. Tang, *A fast two-dimensional median filtering algorithm*, IEEE Transactions on Acoustics, Speech, and Signal Processing, 27 (1979), pp. 13–18.

[2] E. Ataman, V. K. Aatre, and K. M. Wong, *A fast method for real-time median filtering*, IEEE Transactions on Acoustics, Speech, and Signal Processing, 28 (1980), pp. 415–421.

[3] J. T. Astola and T. G. Campbell, *On computation of the running median*, IEEE Transactions on Acoustics, Speech, and Signal Processing, 37 (1989), pp. 572–574.

[4] M. Juhola, J. Katajainen, and T. Raita, *Comparison of algorithms for standard median filtering*, IEEE Transactions on Signal Processing, 39 (1991), pp. 204–208.

[5] J. Gil and M. Werman, *Computing 2-D min, median, and max filters*, IEEE Transactions on Pattern Analysis and Machine Intelligence, 15 (1993), pp. 504–507.
[6] S. L. Tanimoto, *Fast median filtering algorithms for mesh computers*, Pattern Recognition, 28 (1995), pp. 1965–1972.
[7] C.-T. Chen, L.-G. Chen, and J.-H. Hsiao, *VLSI implementation of a selective median filter*, IEEE Transactions on Consumer Electronics, 42 (1996), pp. 33–42.
[8] S.-C. Hsia and W.-C. Hsu, *A parallel median filter with pipelined scheduling for real-time 1D and 2D signal processing*, IEICE Transactions on Fundamentals of Electronics, Communications & Computer Sciences, E83-A (2000), pp. 1396–1404.
[9] R. Roncella, R. Saletti, and G. Savoia, *VLSI design of a parallel architecture 2-D rank order filter*, in Proceedings of the 8th European Signal Processing Conference (1996), Vol. 3, pp. 1483–1486.
[10] C.C. Lin and C.J. Kuo, *Two-dimensional rank-order filter by using max-min sorting network*, IEEE Transactions on Circuits & Systems for Video Technology, 8 (1998), pp. 941–946.
[11] I. Hatirnaz, F.K. Gurkaynak, and Y. Leblebici, *Realization of a programmable rank-order filter architecture using capacitive threshold logic gates*, in Proceedings of the 1999 IEEE International Symposium on Circuits and Systems (1999), Vol. 1, pp. 435–438.
[12] M. Hu, O. Vainio, and D. Gevorkian, *Design of a configurable order statistic filter with FPGAs*, in Proceedings of the 17th IEEE Instrumentation and Measurement Technology Conference (2000), Vol. 2, pp. 1095–1098
[13] B. H. McCormick, *The Illinois pattern recognition computer – ILLIAC III*, IEEE Transactions on Electronic Computers, 12 (1963), pp. 791–813.
[14] M. J. B. Duff, D. M. Watson, T. J. Fountain, and G. K. Shaw, *A cellular logic array for image processing*, Pattern Recognition, 5 (1973), pp. 229–247.
[15] S. F. Reddaway, *The DAP approach*, in Infotech State of the Art Report: Supercomputers, C. R. Jesshope and R. W. Hockney, eds., Infotech, Vol. 2, Maidenhead, England (1979), pp. 311–329.
[16] M. J. B. Duff, *Clip4*, in Special Computer Architectures for Pattern Processing, K. S. Fu and T. Ichikawa, eds., chapter 4, CRC Press, Boca Raton, FL (1982), pp. 65–86.
[17] T. J. Fountain, K. N. Matthews, and M.J.B. Duff, *The CLIP7A image processor*, IEEE Transactions on Pattern Analysis and Machine Intelligence, 10 (1988), pp. 310–319.
[18] K. E. Batcher, *Design of a massively parallel processor*, IEEE Transactions on Computers, 29 (1980), pp. 836–840.
[19] M. J. Little and J. Grinberg, *The 3-D computer: An integrated stack of WSI wafers*, in Wafer Scale Integration, E. E. Swartzlander, ed., Kluwer Academic Publishers, Boston, MA (1989).
[20] E. L. Cloud, *Geometric Arithmetic Parallel Processor: Architecture and implementation*, In Parallel Architectures and Algorithms for Image Understanding, V. K. Prasanna, ed., Academic Press, San Diego, CA (1991).
[21] M. S. Tomassi and R. D. Jackson, *An evolving SIMD architecture approach for a changing image processing environment*, DSP & Multimedia Technology (October 1994), pp. 1–7.
[22] S. Y. Lee and J. K. Aggarwal, *Parallel 2-D convolution on a mesh connected array processor*, IEEE Transactions on Pattern Analysis and Machine Intelligence, 9 (1987), pp. 590–594.
[23] H. Shi, G. X. Ritter, and J.N. Wilson, *An efficient algorithm for image-template product on SIMD mesh-connected computers*, in Proceedings of the 1993 International Conference on Application-Specific Array Processors, Venice, Italy (October 1993).
[24] J. P. Fitch, E. J. Coyle, and N. C. Gallagher, *Median filtering by threshold decomposition*, IEEE Transactions on Acoustics, Speech, and Signal Processing, 32 (1984), pp. 1183–1188.
[25] P. D. Wendt, E. J. Coyle, and N. C. Gallagher, *Stack filters*, IEEE Transactions on Acoustic, Speech, and Signal Processing, 34 (1986), pp. 898–911.

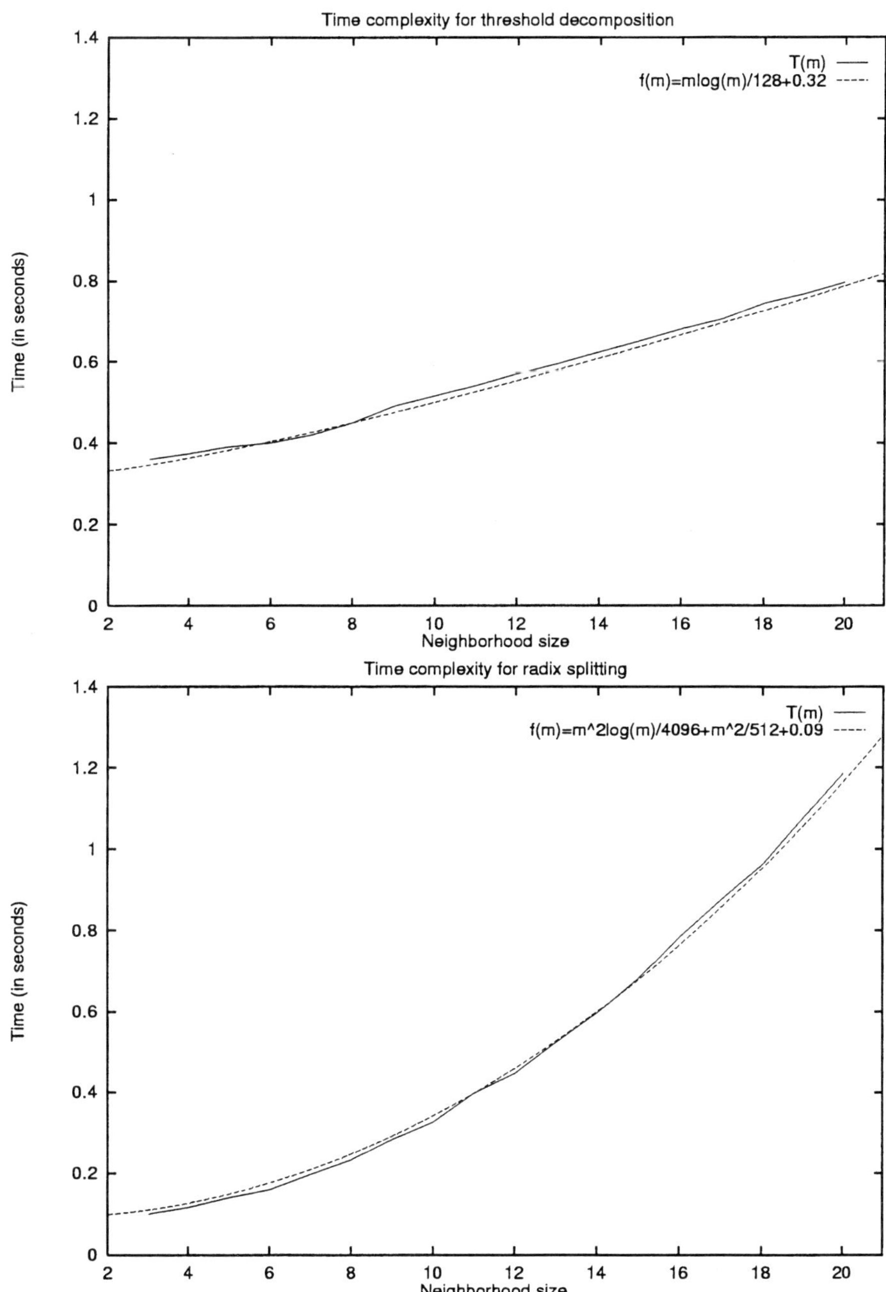

FIG. 5.1. *Time complexity curves for the algorithms*

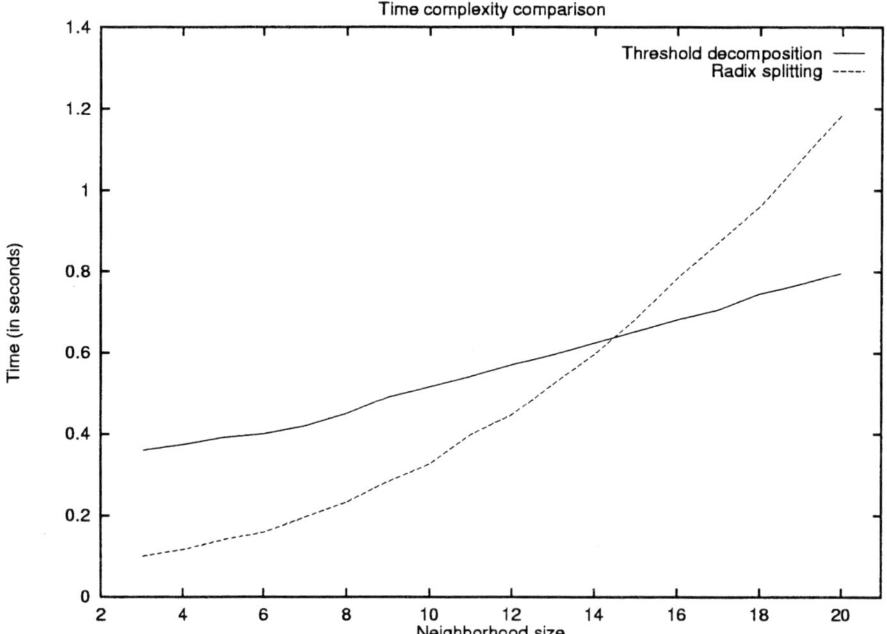

FIG. 5.2. *Comparison of the two algorithms' time complexities*

INDEX

A

B

C

D

E

F

G

H

I

T

U

V

W